Quite Right

QUITE RIGHT

The Story of Mathematics, Measurement, and Money

NORMAN BIGGS

Emeritus Professor of Mathematics
London School of Economics

OXFORD
UNIVERSITY PRESS

OXFORD
UNIVERSITY PRESS

Great Clarendon Street, Oxford, OX2 6DP,
United Kingdom

Oxford University Press is a department of the University of Oxford.
It furthers the University's objective of excellence in research, scholarship,
and education by publishing worldwide. Oxford is a registered trade mark of
Oxford University Press in the UK and in certain other countries

First Edition published in 2016

Impression: 1

Published in the United States of America by Oxford University Press
198 Madison Avenue, New York, NY 10016, United States of America

British Library Cataloguing in Publication Data

Data available

Library of Congress Control Number: 2015944379

ISBN 978-0-19-875335-3

Printed and bound by
CPI Group (UK) Ltd, Croydon, CR0 4YY

PREFACE

My aim is to explain how mathematical ideas evolved in response to the growing levels of organization in human societies, from pre-historic times to the present day. The traditional approach to the history of mathematics focussed on its relationship to the current activities of professional mathematicians, and this resulted in a somewhat narrow view. In recent decades there has been a welcome movement towards looking at history from a broader viewpoint. Combined with new discoveries of artefacts and documents, this approach leads to a better understanding of the part played by mathematics in the human story, and (rather surprisingly) the nature of mathematics itself.

In the spirit of this wider view, the book aims to display the power and beauty of mathematical concepts, which often belie their utilitarian origins. The twin paradigms of logical justification and algorithmic calculation recur throughout. Another recurring theme is the relationship between mathematics and measurement of all kinds, including the measurement of value in monetary terms. So we begin with the measurement of length and area in ancient times, continue with the arithmetical problems of using money in the middle ages, and conclude with the contemporary idea that information itself can be measured.

Many friends and colleagues have contributed in many ways. Perhaps my greatest debt is to those historians of mathematics who set the standard for me: John Fauvel and Jeremy Gray for their splendid creation of the Open University's course MA290, and Jacqueline Stedall for her wise and insightful writings. I thank specifically Bob Burn, Leo Rogers, Adam Ostaszewski, and Robin Wilson, who read draft chapters and provided helpful comments. The participants in the monthly Oxford Forum on the history of mathematics listened patiently to some of my early attempts to tell the story. At the Oxford University Press, Keith Mansfield, Dan Taber, Harriet Konishi, Eilidh McGregor, and Victoria Mortimer provided professional support and advice. Christine and Juliet also provided support and advice, as only family can.

Norman Biggs
October 2015

CONTENTS

CHAPTER 1

The Unwritten Story

I magine that you have to solve your daily problems without the aid of writing or arithmetic. Not so long ago all human beings were in that state. The fact that the human race survived, despite its blissful ignorance, indicates that our prehistoric ancestors were able to cope with problems like dividing the carcass of a slain animal and deciding when to prepare for winter. Such mundane tasks were the seeds from which there emerged the universal problem-solving method that we now call mathematics.

The story of mathematics

This is a story without a clear beginning, and without an end. It tells how some simple ideas were developed, refined, and applied, so that they pervade almost every aspect of life in the twenty-first century.

To make our story intelligible we shall tell it in roughly chronological order. But ideas march to the beat of a different drum, the force of logic, which is often at odds with the arrow of time. Further complication is created by the geographical spread of events: ideas arose independently in different parts of the world, at different times, and were developed in different ways. Globalization is a very recent phenomenon.

We must set aside almost the entire history of our planet, which, we are told, has been in existence for over four billion years. Human-like beings have lived here for only about two million years, and for most of that time they managed to survive without writing down anything about their activities. The written record, what we might properly call history, covers only the last 5000 years or so, and we cannot attach much meaning to the word mathematics before that time. Even if we look for clues a little further back, it is only within the last 30000 years that we can detect even the faintest hints of mathematics, or the related concepts of measurement and money. This means that we are interested almost entirely in the activities of just one species, *Homo sapiens*. To the best of our knowledge there have been no significant evolutionary changes in the mental capacity of our species in the relevant period, despite changes in the environment and consequent improvements in physical stature and longevity. For that reason, we might expect that some ideas that we now think of as mathematical might have appeared, even before we have any written evidence of them. But it must be stressed that in *pre*-historic times we must be content with *pre*-mathematics, a brief account of which will be the subject of this chapter.

Some prehistoric activities, such as cooking and tool-making, had aspects that are inherently linked with pre-mathematical ideas of size and shape. These activities are still with us, but nowadays cooking and tool-making are usually regulated numerically. Our food must be cooked for a certain number of minutes, and our nuts and bolts must have certain sizes so that they fit together. The use of numbers for such purposes originated in the period of human history when concepts began to be expressed in a written language as well as a spoken one. Roughly speaking, this is what we call the Bronze Age, and it was during this period that the basic methods of arithmetic and geometry were first given a definite form. The main part of this book will be devoted to the story of this kind of mathematics.

In due course the effectiveness of mathematical methods led to careful consideration of their foundations, and a coherent body of knowledge was established. The answers provided by mathematics are 'right', and they can be used to resolve issues impartially, without relying on the opinions or brute strength of individuals. When mathematics began to pervade many aspects of human life, both social and commercial, more efficient methods of doing it were discovered. Finally, in the past 500 years, mathematics took on a life of its own, leading to the discovery of powerful new theories, and their application to an ever-widening range of practical problems.[1]

Fairness in the age of the hunter-gatherers

The origins of the notions of fairness and equality are unclear. If we adopt a crude 'survival-of-the-fittest' evolutionary model, then it might be argued that the only primitive concept was inequality. However, a species that settles every dispute by a fight to the death is unlikely to flourish, especially if its child-bearing members are inherently weaker in physical terms.

So it is not unreasonable to suppose that the idea of fairness must have been current in the age when our ancestors lived by hunting animals and gathering plants. Our direct knowledge of this era comes from the archaeological record, and there are a few hints that, in certain circumstances, fairness could have been an issue. For example, there is evidence that the body-parts of woolly mammoths were used for many different purposes. Killing a mammoth is not easy, and it must have been usual for mammoths to be trapped and killed by humans hunting in large groups. For simplicity, let us suppose there were just two families of hunters involved in the kill. How should the carcass be divided between them? The problem is that there are different uses for different parts of the carcass: tusks, skin, fat, meat, bone, and so on, and the two families may have different needs for them. Simply cutting the animal in half may not be the answer.

Fortunately, there is a simple procedure for ensuring that both families will get what they regard as a fair share. It is called *cut-and-choose*, and some readers may recognize it in a more friendly, 'chocolate cake' scenario. First, one family divides the carcass into two heaps, both of equal value according to its needs. Then the other family chooses the heap which is better from its point of view. In this way they are both content. The first family has what it believes to be a half-share, and the second family has what it believes to be a half-share, at least.

Of course, the woolly mammoth scenario is pure speculation, like so much that has been written about the activities of Stone Age men and women. But it is quite possible that the underlying ideas go back a long way. That was the view of the Greek writer Hesiod (*c*.800 BC), who wrote about early mankind in mythological terms. These myths are thought to contain folk-memories of actual events, although they are expressed in the language of fantasy. One of Hesiod's stories tells of a dispute between Zeus, the top god, and Prometheus, a titan.[2] Prometheus tried to trick Zeus by cutting up a bull, dividing the various parts into two heaps, and challenging Zeus to choose one. Zeus saw that the heaps were not of equal value, despite Prometheus's attempts to disguise this, and chose what he considered to be the better one. Readers familiar with Greek mythology will know that Prometheus was punished by the gods, who vented their wrath by subjecting him to cruel torture. If that was not enough, they persuaded Prometheus's brother Epithemeus to accept the fair Pandora as his wife, together with a mysterious box. The opening of Pandora's box was the root cause of all the world's ills, including reality television.

In its simplest form the cut-and-choose method does not require the use of numbers. Nevertheless, it contains the essence of many simple but fundamental ideas. In very recent times the method itself has been studied by mathematicians, and it is now used for international agreements about such things as offshore mineral deposits.[3] Rather more complicated versions have been used for making divorce settlements. The underlying problem of assessing the value of a set of objects involves the concepts of measurement and money, and assessing the value fairly leads to the procedures that we call arithmetic and geometry.

Counting without numbers

It is often claimed that the earliest known mathematical artefacts are objects known as tally-sticks. These objects have lines of cut-marks, clearly made intentionally (Figure 1.1). The ones that have survived are made of bone, but it is likely that most of them were made of wood. The dates of these objects are stated with apparent definiteness in the archaeological literature, but they are 'revised' regularly. It does, however, seem to be agreed that some of them are over 10000 years old.[4]

It is possible to think of many possible uses for such tally-sticks, even in the hands of the hunter-gatherers. Although their precise purpose in that context

Figure 1.1 A tally-stick, made of wood or bone.

cannot be determined, we must avoid explanations that imply levels of mathematical understanding for which there is no evidence whatsoever. The simplest suggestions are usually the best. Although it is reasonably clear that each cut-mark is a record of a single object or event, it is important to stress that this kind of record does not require the notion of the numbers one, two, three, four, and so on. All that is needed is to establish a correspondence between the marks on the stick and the objects themselves. One possibility is that the marks might represent a set of stone tools, such as hand-axes. If each cut-mark corresponded to a tool, the tally-stick could be used to check that no tools had been lost, as in Figure 1.2. The difference between fifteen items and fourteen is not easily recognized at a glance, and nowadays we would check by counting the items, using our names for the sequence of numbers. The prehistoric tally-stick provided a method of checking that did not require names or signs for numbers. The names, and the rules for manipulating the signs that we now call arithmetic, belong to a much later stage of development.

In recent times mathematicians decided that they should examine the logical basis of the number-system. In due course it turned out that the concept of a one-to-one correspondence, such as that between the tools and the marks on the tally-stick, is fundamental. At first sight this reversion to prehistoric notions might

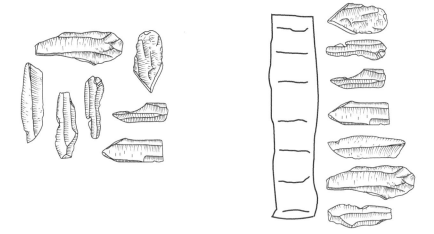

Figure 1.2 A set of stone tools and how a tally-stick could have been used to check them.

appear surprising, but on reflection it offers some profound insights into the relationship between abstract ideas and human understanding.

Around 10000 BC there was a significant change in the climate of the northern hemisphere, usually referred to as the end of the Ice Age. Before that time men and women in all parts of the world led very basic lives. Most of the concepts and practices described in this chapter would have been familiar to them, but only in simple forms. They would have understood the idea of sharing fairly, and they would have used the tally method for keeping records. After the ice retreated there followed a transitional period of at least 6000 years, during which the way of life of people in many parts of the world, not only those who had been under the ice, changed dramatically. We have no written records to reveal the cause of the change, but the archaeological evidence is clear: people settled in one place, and learned how to cultivate crops and keep animals. Later, among the written records of the people who lived in Mesopotamia, the land between the rivers Tigris and Euphrates, there are mythological accounts of this transition, not unlike the Greek myths that are more familiar to us, but written well over 1000 years earlier. One such myth is known as 'The Debate between Grain and Sheep'. It contains fantastic accounts of the pros and cons of the two basic modes of agriculture, with a few insights into practical details.[5] For example, when the Grain-farmer challenges the Shepherd about his comfortable life, he says

> every night your count is made and your tally-stick put into the ground, so your herdsman can tell people how many ewes there are and how many young lambs, and how many goats and how many young kids.

The wording suggests that the tally-stick had acquired a more sophisticated function by this time. As we shall see in the Chapter 2, there is indeed ample evidence that forms of arithmetic were being used in Mesopotamia by about 3000 BC, well before the mythical Debate was written down. However, the transition took place much more slowly in other places. Even now there are primitive tribes in the transitional stage, a few of them still living like hunter-gatherers. Throughout that stage fundamental concepts were developed gradually, in response to the needs of a more organized society. People began to use words for specific numbers, such as their *two* hands and the *five* fingers on each hand. But theirs was a spoken language, not a written one, and we have no idea what those words were.

The origins of money

Another feature of everyday life in prehistoric times would have been the exchange of goods in a variety of social situations. Among the hunter-gatherers the acquisition of food had been uncertain and irregular. One group might catch more fish than they needed, but be short of fruit. Another group might be in the opposite situation. In this case an exchange would have been the obvious alternative to the risky business of a winner-takes-all fight. Thus there can be little doubt that barter was common. There is, however, an obvious problem with barter as a means of satisfying the needs of all parties. A group with too much fish and not enough fruit may not be able to interact directly with another group that has too much fruit and not enough fish. More generally, any group that wishes to take part in an exchange may find itself involved in a series of complex negotiations with several other groups to satisfy its needs. That is the problem of 'the inconvenience of barter'. In the transitional period people began to use a commodity that can be exchanged for fish or fruit or any other commodity, and so they avoided the problem of matching one group's needs with another's. This special commodity took many forms, but we now call them all by the name *money*. So it might be argued that the origin of money lay in the inconvenience of barter. Indeed that was the opinion of the Greek philosopher Aristotle; in his *Ethics*, written around 350 BC, he claimed that

> all things or services then, which are to be exchanged must be in some way reducible to a common measure. For this purpose money was invented, and serves as a medium of exchange: for by it we can measure everything.

Aristotle's opinion was echoed faithfully for a very long time: for example by Nicholas Oresme in his book *De Moneta* (around 1360 AD), and by Stanley Jevons in his book *Money and the Mechanism of Exchange* (1872). It became the accepted wisdom among economists. But in the twentieth century a different theory emerged. It was the work of anthropologists who studied the primitive tribes that had managed to survive without coming into contact with civilization. Of course, we cannot surely identify the practices of these people with those of the hunter-gatherers, but the parallels are very suggestive. What the anthropologists found was that the objects used as primitive money, for barter and exchange, were also used in other ways. Furthermore, these other ways appeared to be more fundamental, in that they were part of essential social functions, rather than trade. For example, the acquisition of a bride often entailed a payment to the bride's father, in recognition of him having lost the services of a daughter, who had become a means of ensuring the survival of the species. Payments were also common as compensation for injury or death, and for protection from powerful and predatory neighbours.[6]

It is important to stress that the objects used as primitive money were not coins as we know them, and (with few exceptions) they were not made of precious metal. The use of coins as a form of money comes later in our story, and before that many objects served as money: some of them, such as the large stones used until recently on the island of Yap, were spectacularly inefficient. In the mainstream of progress towards a modern economic system, the most significant forms of money were those that also had an intrinsic value, such as cattle and grain. From the days of the earliest farmers these items were widely available. They could be used directly as part of a barter-exchange, or they could be transferred between farmers in a series of exchanges, so that the wants of all could be satisfied. Clearly, these transactions stimulated the development of arithmetic (how many oxen will buy me a bride?)

Figure 1.3 Cowrie shells, each one is about 1 cm long.

and measurement (how much grain will buy me an ox?). In Chapter 2, we shall see how these ideas had developed by the time of the earliest written records.

But we cannot leave the subject of money here without a passing reference to one of its most widespread forms, the cowrie shell (Figure 1.3). The cowrie is a small mollusc whose shell has many of the properties required for a money-object. It is durable, easily portable, and can be obtained in large quantities. In fact, the cowrie is the natural ancestor of the kind of token coinage that we use today, having little or no intrinsic value. We do not know whether cowries were used in the dark age before written records, but they are referred to in Chinese records from at least 3000 years ago. More recently, even in the nineteenth and twentieth centuries, many thousands of tons of cowries were being dredged from the seas around the Maldive Islands, to be used as money in Africa and India.

In summary, money originated in prehistoric times because it was useful in several different kinds of transaction, both social and commercial. When the hunter-gatherers became farmers they used common items such as cattle and grain to pay for the acquisition of all the goods and services that they needed. This form of primitive economy inevitably led to the development of more advanced ideas, which we now know as arithmetic.

Mysterious moonshine

It seems reasonable to suppose that, from their earliest beginnings, human-like beings were consciously aware of the annual pattern of the seasons. For them the seasons would have been defined by the observable changes in the environment: Spring is when the trees and plants sprout new growth, Autumn is when the leaves fall.

The early humans must also have been aware of the daily sunrise and sunset, and the monthly waxing and waning of the Moon. However, these regular events were regarded as great mysteries, the work of supernatural forces. For almost all of human history there was no understanding of how the Sun, the Moon, and the Earth moved in relation to each other, let alone the physical causes of their motion. Complicated belief-systems were constructed to explain, for example, why the rising of the Sun is so inexorable, and people who could make a good story out of such things became powerful as priests and gurus, styling themselves 'wise' men and women. It was probably these wise-ones who first began to work out that there are links between the days, months, and years. But the numerical language for expressing such links was unknown, so there could be nothing like our modern calendar with its numerical basis, such as the 365 days to a year. Before the coming of writing and arithmetic, a calendar could exist only in what we might call an analogue form. For example, some of the early tally-sticks could have been used to check the days from one full Moon to the next. The fact that the survival of the species depends on a menstrual cycle of roughly the same length as a Moon-month is another factor that could have been recorded in the same way.

The wise-ones would have been very powerful, especially when they discovered how to work out the relationships between the days and the seasons. That knowledge, even in rudimentary form, allowed for predictions, which suggested that they had magical powers over the environment. This apparently happened in the aftermath of the Ice Age, before there is any written record. At first they could have used simple artefacts: the shadow cast by an upright stick indicates, not only the passing of the daylight hours, but also the seasons of the year, by virtue of its length. But primitive instruments made of wood are not durable and, even before the dawn of civilization, stone artefacts would have been preferred. By the time civilization emerged in Egypt, the stones had become massive obelisks, on which all kinds of useful data could be recorded. In Britain there are circles of standing stones, some of them replacing earlier constructions of wood. The ruinous state of these ancient monuments makes it impossible to understand exactly how they were used, but it is quite possible that they were designed by the wise-ones to display their insights into the clockwork of the seasons. Because these matters were of vital importance to the community, the monuments were also the places for ceremonial activities, such as thanking the gods and offering gifts to ensure their continuing goodwill. The ceremonies were doubtless organized by the local chiefs and their advisers, who could use the occasion to emphasize their personal power.

One of the most serious defects of twentieth-century writing on the history of ideas was the tendency to attribute knowledge of sophisticated concepts to prehistoric peoples. Unfortunately, these theories were advanced by people who were, in some respects, the 'wise-ones' of our age, and they are still thoughtlessly repeated by others who ought to know better. For example, it was noticed that some of the groups of marks on a bone tally (like the one shown in Figure 1.1) represent special numbers that we call prime numbers. This led to the suggestion that the person who marked the bone was familiar with a complex range of arithmetical ideas that did not (and could not) exist at that time. Another example was the theory that the stone circles had been constructed using a 'megalithic yard', the magnitude of which had been standardized across a large part of Europe and Asia (it was claimed to be about 83 centimetres, in modern terms). In fact archaeological evidence tells us that the inhabitants of Britain at the relevant time (*c*.3000–2500 BC) were living in a prehistoric culture, and there is certainly no evidence to suggest that they were familiar with complex systems of measurement. To them, the obvious measure of length would have been a human stride, and it is easy to imagine how that might have been used by the builders of the circles. As a stride is roughly 83 centimetres, it is not surprising that some evidence may appear to support the megalithic yard theory, even though common sense rejects it.[7]

The rest of this book is devoted to tracing how the tenuous prehistoric notions of measurement and money developed into concepts that we now take for granted, and without which modern life would be impossible. The thread that holds this story together is what we call mathematics.

CHAPTER 2

The Dawn of Civilization

A bout 4500 years ago people in some parts of the world were living in large settlements. If you belonged to one of these settlements you would notice that some aspects of life were regulated by strange signs and symbols. The most mysterious symbols were those that represented numbers, and the few people who understood these symbols were important members of the community. If you were one of them, you would be able to use numerical methods to solve practical problems, like distributing the store of grain fairly.

Writing and counting

How many sheep do I have? The 'Debate between Grain and Sheep', mentioned in Chapter 1, tells us that the early farmers of Mesopotamia kept track of their flocks by the already ancient method of making marks on a tally-stick. There is some archaeological evidence suggesting that better methods may have been in use as early as 8000 BC, but there is no general agreement on the details. We only know that by about 3000 BC, when the written records begin, very significant developments had taken place. Slowly, but inevitably, words and symbols for numbers had emerged and, for the people who could understand them, they were an important part of their lives. To put this process in perspective, it is worth remarking that the 5000 years separating the conjectured origins from the first written records is roughly the same length of time as that which separates the first written records from the present day.

By about 3200 BC a large settlement had been established at Uruk in Mesopotamia. Many of the inhabitants were not directly involved in farming, and a clear social hierarchy was in operation. A few of the higher-ranking citizens controlled the economic life of the city by administering the distribution of resources, so that the building workers got the food they needed, for example. Archaeologists have found many small clay tablets that were used to record these administrative mechanisms. The tablets have been inscribed by making marks in the wet clay: some of the marks are pictograms representing commodities of various kinds, and others

Figure 2.1 Pictograms and number-signs used on clay tablets from Uruk, c.3150 BC.

are symbols representing numbers (Figure 2.1). Many such clay tablets were apparently discarded and re-cycled for use as building rubble. When dry they are almost indestructible, and this provides us with (currently) the best evidence of how the use of numbers was developing around the end of the fourth millennium BC.

The first important revelation provided by the Uruk tablets is that numbers were being represented in abbreviated form. The abbreviations differed according to what was being counted, but the idea was simple enough. For example, if the name for five sheep was a 'batch', then a batch could be represented by one symbol, and twenty-three sheep would not be denoted by twenty-three single marks, but rather by four batch-symbols and three single marks. A system of this kind has the unfortunate consequence that twenty-three sheep and twenty-three baskets of grain might well be written quite differently, because the corresponding abbreviation for grain could be a 'stack' of six baskets, so that twenty-three baskets were represented by three stack-symbols and five single marks. We ought not to scorn this diversity, for it was not very long ago that some children had to be taught that twelve inches make a foot, but eight pints make a gallon. Fortunately for the people of Uruk their system was not frozen in time, and improvements could be made.

Contemporary with the Uruk tablets, there is evidence of the use of number-signs in Egypt, in particular from Naqada, a settlement on the Nile about 600 km south of Cairo. Many small ivory labels have been found there, each with a hieroglyphic inscription on one side, and a number on the other side. Here too the numbers are expressed in an abbreviated notation, with a symbol for units, a symbol for tens, a symbol for hundreds, and so on (Figure 2.2). The function of these ivory labels is not fully understood. But whatever their purpose, there are enough examples to justify the assertion that an effective system of representing numbers was in use in parts of Egypt at that time.

In summary, the evidence suggests that the development of the art of writing was closely linked to the use of symbols for representing numbers.[1] Both written text and number-symbols arose in response to the need for keeping account

Figure 2.2 Symbols on ivory labels from Egypt, c.3150 BC. They are believed to represent one hundred, four tens, and three units.

of the various things that were important in the early agrarian economies. The number-symbols were the original tools of mathematics, and as such they played an important part in the process we call civilization.

Operating with numbers

The clay tablets used in Uruk at the end of the fourth millennium BC were the first in a long series of similar records. The marks on the tablets gradually evolved into a style of writing known as cuneiform and, as we shall explain shortly, conventional signs for numbers were also used. This uniform system of writing was in use throughout a large part of what is now southern Iraq, around 2000–1500 BC. We shall refer to this period and its culture as the Old Babylonian (although the association with Babylon itself is only nominal).[2]

By that time the economic organization of society had become more complex, and correspondingly sophisticated ways of dealing with numbers had been developed. The procedures were based on the number sixty. For example, as the number we write as 75 is equal to *one* sixty and *fifteen* units, it was represented by combining the signs for *one* and *fifteen*. For larger numbers, the units were sixty-sixties (our 60 × 60 = 3600), sixty-sixty-sixties (our 60 × 60 × 60 = 216000), and so on. This is known as a *sexagesimal* system, from the Latin word for sixty.

Because the basic unit was 60, individual signs for the numbers from 1 to 59 were needed. They were formed from the cuneiform symbols for one and ten, as shown in Figure 2.3. For example, our 17, made up of one 10 and seven 1s, was denoted by the ten-symbol and seven one-symbols. Numbers larger than 59 were written in sexagesimal form, so that the number 257 was written as four one-symbols (denoting four sixties) and the combined seventeen-symbol.

The fact that these numbers were written on clay tablets, so many of which have survived, provides us with splendid evidence of how mathematics was being used in the Old Babylonian period. Although the sexagesimal system itself is no longer with us, it is remarkable that the number 60 still plays a major part in the management of our daily lives. If you have ever wondered why each hour has 60 minutes, you can blame the Old Babylonians. The historical reason for the success of the system was that it could be used to carry out the numerical operations

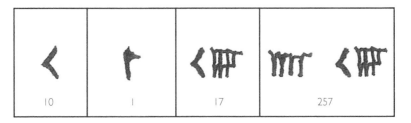

Figure 2.3 Symbols for numbers used on cuneiform tablets

required by the administrators who controlled the economic life of the region. These operations are what we call arithmetic. Essentially they are a clever form of juggling with number-signs, so that the answers to certain practical questions can be obtained.

To help us understand the methods used in Mesopotamia, historians have devised a way of translating sexagesimal numbers into our modern form, without destroying the basic features of the system. For example, the symbol for 75 is translated as [1 | 15], and the symbol for 257 is translated as [4 | 17]. We shall call this the 'mixed notation'.

The advantage of the mixed notation is that we can use it to explain clearly how the elementary arithmetical procedures were done. The simplest of these is addition. If I have a flock of [1 | 15] sheep and I acquire my neighbour's flock of [2 | 23] sheep, how many sheep will I have? The obvious method is to count the sixties and the ones separately: so I will have 1 + 2 = 3 sixties, and 15 + 23 = 38 ones. In mixed notation, the answer is [3 | 38] sheep. If the total number of ones had exceeded 59, then the number of sixties would have been increased by 1 and the excess left in the ones: for example [1 | 45] sheep and [2 | 39] sheep make [4 | 24] sheep. Observe that operating with numbers in sexagesimal form requires knowing all the answers for numbers from 1 to 59, and the scribes had to learn these facts first. For example, to solve the preceding problem they would have to know that 45 and 39 make [1 | 24].

Another level of complexity arises when we come to multiplication. Suppose the problem is to multiply [5 | 7] and [3 | 2]; that is,

[5 sixties and 7 ones] times [3 sixties and 2 ones].

The method was based on splitting up the numbers in the following way. First, the 5 sixties are multiplied by the 3 sixties, giving 5×3 sixty-sixties. Then we have 5×2 sixties and 7×3 sixties. Finally, there are 7×2 ones. Working out the numbers 5×3 and so on, and collecting up the terms, gives the answer 15 sixty-sixties, 31 sixties, and 14 ones. In the mixed notation the result is:

[5 | 7] times [3 | 2] is equal to [15 | 31 | 14].

The procedure for multiplication implicitly involves rules about splitting up the numbers and collecting similar terms. Mesopotamian scribes were presumably taught to regard these rules as self-evident, based on an innate sense of how numbers represent reality. It was many thousands of years before mathematicians began to question whether they are truly universal. Another significant point is that the scribes had to learn a lot more 'tables' than we do. Because our modern system of numeration is based on 10 rather than 60, we only need times-tables for the numbers from 1 to 9, whereas the Mesopotamians had to cope with 1 to 59.

We are fortunate to have some evidence of how the scribes were trained in the Old Babylonian period. In 1952 archaeologists excavating at Nippur in southern

Iraq located a building that had evidently been used as a kind of school. The floors and walls of this building contained well over 1000 clay tablets, obviously discarded and re-used for a more basic purpose. By studying the style and attributes of these tablets in detail, it became possible to work out how the curriculum was organized. The scribes had much to learn. They had to master the art of writing and the rules for multiplication, as well as extensive lists of weights and measures. It is quite possible that they did not have to remember all these facts, because tables for reference were available. But even learning how to use the tables would have required a high level of mathematical understanding. This was not mass education in the modern style, but the training of an elite who would become rich and powerful. And what was perhaps the most difficult part of their work has still to be mentioned.

Fair division in the age of arithmetic

The problem that gave the Mesopotamian administrators the most trouble was the sharing of resources. The rise of arithmetic offered the prospect of fair division based on numerical methods, but there were obvious difficulties. When twenty loaves had to be divided fairly among five people, they each got four, and everyone was content. But if there were nine people, then the division could not be done exactly: each person got two loaves, but there were two left over. The two left-over loaves had to be divided into nine equal parts in some way, so that everyone got the fraction that we call two-ninths. There are several ways of looking at what we call fractions, and we must be careful not to confuse the rules that we ourselves were taught as children with those used 4000 years ago. The ancient civilizations of Mesopotamia and Egypt both discovered ways of dealing with fractions, but they were different ways, and our way is different again. To emphasize this distinction, we shall frequently use the term 'fractional part' in the following discussion.

The Mesopotamian method for working with fractional parts was an extension of their sexagesimal notation. For whole numbers they used sixties and larger units, and so for fractional parts they used sixtieths and the corresponding smaller units. The same notation could be used in both cases, so the symbol that we translate as [4 | 17], and previously interpreted as 257, could also stand for

$$\frac{4}{60} + \frac{17}{3600} = \frac{257}{3600}.$$

This may appear confusing, but it was based on a brilliant insight: fractional parts can be manipulated in the same way as whole numbers. Hence they can be used to great effect in arithmetical calculations. In practice the meaning of an ancient calculation would be clear from the context; our difficulty is that 4000 years later the context itself is not always clear.

To solve division problems with the sexagesimal notation, the method was based on another table. In modern terms, the idea was that dividing by nine (say) is the same as multiplying by the fraction one-ninth. Thus the scribes were trained to use a table that contained a list of entries like the following:

the 8th-part is [7 | 30];
the 9th-part is [6 | 40];
the 10th-part is [6].

Here the sexagesimal notation for fractional parts came into its own, because the rules for multiplication were the same, whether [6 | 40] meant 6 sixties and 40 ones, or what we would write as

$$\frac{6}{60} + \frac{40}{3600},$$

which (as you can easily verify) is equal to our one-ninth.

As an example of how easy it was to apply these rules, let us pretend to be a bureaucrat who has to solve the problem of dividing 2 loaves equally among 9 people. First, we refer to our table to find that one 9th-part is [6 | 40]. Then we deduce by simple addition that two 9th-parts make [13 | 20]. Remarkably, this is the answer: the numbers 13 and 20 are all that we need to share out the loaves, as shown in Figure 2.4. We must assume that each loaf can be divided into 60 parts (slices). Suppose 13 of them are given to each of the 9 people. This accounts for $9 \times 13 = 117$ of the 120 slices, with 3 left over. Each of these 3 slices is then subdivided again into 60 parts, making 180 slivers. If 20 of these slivers are given to each person then the allocation is done, as 9×20 is 180. In modern notation, the result is that

$$9 \times \frac{13}{60} + 9 \times \frac{20}{3600} = 2.$$

It cannot be claimed that this method is entirely practical in the case of the loaves, but if the objects in question were baskets of grain, the operations could be carried out fairly easily, using suitable vessels. The significance is that it is a universal procedure which can, in theory, be applied to any division problem. It requires a modest degree of competence to use it properly, but it is not unduly demanding of time or effort. The unknown man or woman who discovered it deserves the title of genius, for this division-arithmetic was the first major discovery in the history of mathematics. Not only was it of great importance in the administration of the

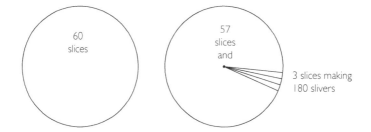

Figure 2.4 Dividing 2 loaves among 9 people, each one getting 13 slices and 20 slivers.

necessities of life, it allowed the Mesopotamians to solve complicated problems about the physical world, such as the relationship between the cycles of the days and the years. By this means they were able to establish a proper calendar.

Enough has been said already to justify the claim that division is the most complicated of the basic arithmetical operations. The Mesopotamian method was just one way of dealing with it. A rather different approach was used in Egypt, where arithmetic had also reached an advanced stage in the second millennium BC. Although much has been written about Egyptian arithmetic, it is important to remember that the documentary sources are very few, and almost all are unprovenanced. In contrast with the Mesopotamian situation, only a handful of Egyptian mathematical texts from this period survive, the most famous being the one known as the Rhind Papyrus, which dates from around 1650 BC. Historians disagree about the original purpose of this document, but we can nevertheless learn a lot from it.[3] One point that is very clear is that the Egyptians had developed systems of arithmetic using number-symbols like those shown on the ivory labels from Naqada (Figure 2.2). The procedures for addition and multiplication were well established, but the details varied from problem to problem, suggesting that the scribes themselves had a good understanding of the basic principles, and could use whatever techniques seemed appropriate in each case. When it came to division, they used a clever technique that still interests mathematicians today, although this modern interest is theoretical, rather than practical.

According to the Rhind Papyrus, the Egyptian approach to division-arithmetic was based on what we would now regard as a very special kind of fraction. But our viewpoint does not necessarily explain how the Egyptians discovered their method and made it work. Here is an example. Suppose a bureaucrat using the Rhind Papyrus was faced with the problem discussed above, dividing 2 left-over loaves equally among 9 people. He or she would begin by referring to one of the tables in the papyrus, where the relevant entry is something like:

to divide 2 among 9 use 6th-parts and 18th-parts.

Using this rule the bureaucrat would first divide the loaves into 6th-parts, and give one to each person. Three 6th-parts would remain, and these would be split into thirds, making nine 18th-parts of a loaf. Each person would get one of them, completing the allocation, as in Figure 2.5.

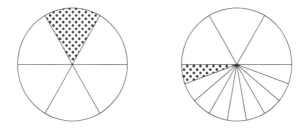

Figure 2.5 The old Egyptian method of dividing 2 loaves among 9 people.

In modern notation, the rule is equivalent to:

$$\frac{2}{9} = \frac{1}{6} + \frac{1}{18},$$

where the required fraction two-ninths is decomposed into a sum of fractions with the number 1 on top.[4] Such fractions are now known as *Egyptian fractions*, although the name is something of a misnomer, because the Egyptians did not think of them as fractions in our sense. Using modern methods it is fairly easy to show that any fraction can be decomposed into Egyptian fractions, but the Egyptians did not have the advantage of our notation and they were content with making a list. In fact the decomposition is not unique: but the Rhind table contains only one such decomposition for each fraction, and nobody has been able to give a satisfactory explanation of how the choice was made. This suggests that the list was originally compiled by highly skilled arithmeticians, probably using a range of trial-and-error techniques. Less skilled scribes would have been content to use the table as given, without attempting to improve it.

Our discussion of ancient arithmetic must end with a warning. In some ways the historian of mathematics has an easy task when faced with the problem of translating ancient arithmetical texts. Although the methods used are often unfamiliar, the universal consistency of arithmetic itself is usually a great help in deciding what is being said. However, some difficulties remain. Merely translating a calculation into modern notation does not necessarily help us to understand its purpose. For this reason it is wise to think carefully before claiming that the ancients knew about certain parts of advanced mathematics. Their interest in arithmetic was entirely practical, and there is no reason to suppose that they took an interest in the abstract properties of numbers, which we now call the Theory of Numbers. That had to wait for a new kind of intellectual approach, which we shall describe in Chapter 3.

Measuring length and area

Measurement is the assignment of a number to an object to describe one of its characteristic properties. Nowadays the measurement is expressed as a number of standard units: so my height is 185 centimetres, and this number of centimetres is counted out on a tape-measure. It is reasonable to assume that simple measurements of length were being made in this way long before we have any written record of them. Indeed, this form of measurement appears to be closely linked to the notion of counting itself, and the evolution of words for numbers. Originally the units were determined by parts of the human body; for example, the length of a human forearm (about 50 centimetres) would have been a useful unit for domestic purposes, such as carpentry and building houses. When written records began, a unit of roughly this size is mentioned in many of them. The actual name varies from place to place, but we call it by the Latin name *cubit*, which is of course much later. In ancient Egypt the cubit appears in the Rhind Papyrus, and is denoted by a hieroglyph representing a forearm.

To measure length more exactly, smaller units such as *palms* and *fingers* were used, also based on human anatomy. However, some of the earliest written evidence is related to larger units, used for measurements of a different kind. This is concerned with measuring the area of a plot of land, a concept which could have arisen back in the age of the hunter-gatherers, when tribal groups claimed to control specific territory, defined by natural boundaries such as forests and rivers. The advent of settled farming was inevitably accompanied by the notion that certain pieces of land belonged to certain family groups. This is a more specific form of ownership and, in due course, it led to problems that could only be solved by mathematical methods. How much land do I have? If I wish to divide it equally between my two children, how should I do it? Such were the original problems of geometry, literally earth-measurement.

Currently the world's oldest known piece of recorded mathematics is a tablet found in Uruk and dating from around 3200 BC. The tablet is known as W.19408.76 and it is kept in the Iraq Museum.[5] Its subject is the size of a field. The tablet itself has no diagram but, from the content, it seems likely that the field was bounded by four straight lines (Figure 2.6). The writing on the tablet is not explicit, but it seems that the scribe tried to estimate the size of the field by multiplying the average length of the sides AB and CD by the average length of the sides AD and BC. This method was commonly used in later times, but of course it is only approximate, and the numbers used in the calculation also seem to be crude approximations. The significant fact is that the result is obtained by multiplying two lengths. Thus, for the scribe in Uruk, and for us, area is a two-dimensional concept. An area-measure is defined as the product of two length-measures, and is therefore expressed in units like the *square metre*. For example, a rectangular plot of land with sides of length 7 metres and 4 metres has an area of 7 × 4 square metres (Figure 2.7). In a simple case like this the square metres are quite obvious, and we can work out that the answer is 28 simply by counting the squares.

When the author of tablet W.19408.76 found the area of the field, the unit of length was a *rod*, roughly equivalent to 6 metres. The sides of the field could have been measured with a rope of this length, a method that is specifically referred to in later documents. It must also be noted that the Uruk tablet was written at a time

Figure 2.6 How can we measure the size of this field?

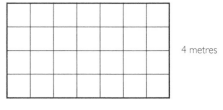

Figure 2.7 The area of the rectangle is 7 × 4 = 28 square metres.

4 metres

7 metres

when multiplication could only be done in very simple cases, although 1000 years later it would have been a routine task for scribes of the Old Babylonian period. A more fundamental difficulty is that most fields have rather irregular shapes, and so the area units, be they square rods or square metres, cannot be laid out as they are in a rectangle. When we try to cover an irregular field with a grid of squares there will be many squares that lie partly inside and partly outside the field, and there is no easy way of accounting for them. At some point in the third millennium BC this difficulty was resolved by the discovery of one of the most important results in the whole of geometry. Suppose we have a triangular field, such as the one shown (T) in Figure 2.8. Imagine another triangle of the same size and shape placed adjacent to the first, as in the second diagram. Now slice off a piece from the right-hand end and replace it on the left. The result is a rectangle (R), clearly equal in area to two of the given triangles. The horizontal side of this rectangle is the *base* of the triangle. Its vertical side is the *height* of the triangle, the length of the perpendicular line from the base to the top vertex. So it follows that the area of the triangle is:

half the base times the height.

This is one of the most useful results in the whole of mathematics. The difficulty of dividing our triangular field into little square pieces has been overcome by some very simple imaginary operations. The importance of the result is dramatically increased when we realize that any figure bounded by straight lines can be divided into triangles, and hence its area can be measured exactly. For example the area of a four-sided field like that previously mentioned can be calculated exactly by measuring the length of a diagonal and the heights of the two triangles into which the diagonal divides the field. This process of *triangulation* was to become the basis of the art of surveying, as illustrated (Figure 2.9) in John Cullyer's *Gentleman's and Farmer's Assistant* (1813).

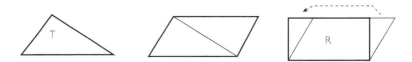

Figure 2.8 The area of the triangle (T) is half the area of the rectangle (R).

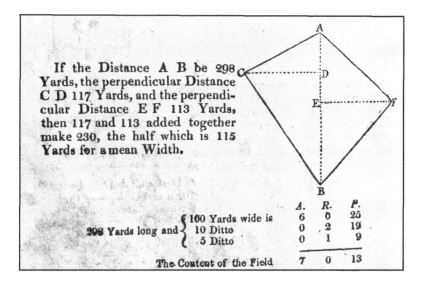

If the Distance A B be 298 Yards, the perpendicular Distance C D 117 Yards, and the perpendicular Distance E F 113 Yards, then 117 and 113 added together make 230, the half which is 115 Yards for a mean Width.

298 Yards long and { 100 Yards wide is

10 Ditto

.5 Ditto

The Content of the Field

	A.	R.	P.
	6	0	25
	0	2	19
	0	1	9
	7	0	13

Figure 2.9 Measuring a field by triangulation, from an early nineteenth-century book.

Measuring quantities

In the Old Babylonian period the economic organization of the larger settlements depended on storing large amounts of grain and distributing it to the inhabitants. Records on clay tablets suggest that the grain was stored in pits of a standard size, one rod (about 6 metres) square and one cubit (about 50 centimetres) deep (Figure 2.10). The resulting measure was known as a *volume-sar*.

To distribute the grain to the people, a different kind of volume-measure was used. It was known as a *sila*, and it was determined by the capacity of a suitable vessel. A sila was very roughly equivalent to a litre in modern terms, so the vessel was similar in size to those in which drinks are now sold. A very interesting clay tablet from the Old Babylonian period deals with the problem of making a sila-vessel, as well as providing several remarkable insights into the mathematical achievements of the time.[6] Suppose we want to make a vessel with the capacity of one sila, in the form of a cylinder with a circular base. The problem is: given the

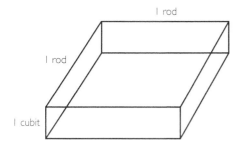

I rod

I rod

I cubit

Figure 2.10 The volume-sar: 1 rod × 1 rod × 1 cubit.

diameter of the base, how high should the vessel be? The solution requires several pieces of data, including the fact that one volume-sar contains 21600 sila, and the relationships between the units of length that were in use at that time:

$$30 \text{ fingers} = 1 \text{ cubit}, \quad 12 \text{ cubits} = 1 \text{ rod}.$$

Given these rules, it follows by arithmetic that one sila is the same as 180 'cubic-fingers'.[7] As $180 = 6 \times 6 \times 5$, a sila could have been measured with a vessel having a square base with sides of length 6 fingers and a height of 5 fingers (Figure 2.11).

The problem is to determine the height of a sila-measure in cylindrical form, if the diameter of the circular base is 6 fingers. First, we must find the area of the circular base: counting little squares does not work with a circle. In the Old Babylonian period such calculations had to be done by approximate rules, and we must be careful not to confuse these rules with our modern formulas involving the number we denote by π (the Greek letter pi). The rule stated on this tablet is to multiply the diameter by three, then multiply the result by itself, and then take the twelfth part; so the area of the base in 'square fingers' is:

$$(3 \times 6) \times (3 \times 6) \times \frac{1}{12}.$$

This works out as 27. Now the rest is easy. We know that one sila is 180 cubic fingers, so the height of the vessel must be 180 divided by 27, or six and two-thirds fingers.

There is another, quite different, way of measuring quantities. It is called *weight* and is based on the simple fact that objects possess a mysterious property that makes them difficult to lift and move. It must have been clear to the early farmers that different materials possess this property in different degrees: a bucket full of water has less weight than the same bucket full of sand, so weight is not simply another way of measuring volume. We must also remember that the physical foundations of the concept of weight were not clarified until the seventeenth century AD, and so we must avoid using the words 'gravity' and 'mass'. The ancient idea of weight was based on experience, rather than theory. Because the weight of an object is not immediately apparent to the human eye (unlike length and volume), the measurement of weight had to be done by a mechanical device which could produce clearly visible results.

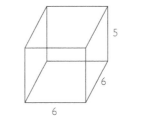 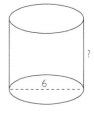

Figure 2.11 Two sila-measures. What should be the height of the second one?

The simplest mechanism for measuring weight probably originated from the common practice of carrying loads using a pole across the shoulders, with (for example) a bucket of water suspended from each end. This mechanism is an excellent means of deciding whether two objects have the same weight, and it is very likely that it inspired the earliest form of weighing machine, the equal-arm balance. Many illustrations of weighing have been found at ancient Egyptian sites.[8] The one shown in Figure 2.12 depicts two people using a balance of this type. The actions of the participants (and the hieroglyphic text) show clearly that the weighing has led to agreement, because the level beam is visible proof of correctness.

It is possible to distinguish several stages in the practical measurement of weight, and it must be remembered that progress was made at different times in different places. The first stage is the very simplest instance of the problem of fair division. Using the equal-arm balance, it is easy to divide a quantity of material into two equal parts, in a way that is both simple and visibly correct. However, the next stage, division into more than two parts, requires slightly more work. If a number of people are to receive equal shares of some commodity, this must be done by balancing each share against the same object. The reference object could simply be one of the proposed shares, but it might be more convenient to use a different kind of object, such as a stone. The stone itself then becomes what we usually call, rather confusingly, a 'weight'. (For clarity we shall sometimes use the term 'weight-object'.) This stage does not represent measurement in the full sense, because no numerical value is assigned, so it could have been used in societies where arithmetic was not yet known. For example, in the course of an extensive excavation of an Iron Age settlement at Danebury in Britain, archaeologists found 68 weight-objects made of stone, many of them fitted with an iron ring for ease of handling. The sizes of these weights display no discernible pattern, and it may therefore be inferred that the inhabitants of Danebury at that time had not progressed beyond the non-numerical stage of weighing. As we shall see, in other parts of the world the measurement of weight had become more closely associated with numerical expertise.

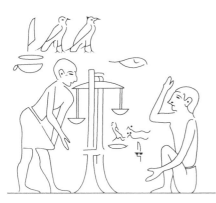

Figure 2.12 The level beam is visible proof that the two loads are equal.

The origins of units of measurement

The unit we call a cubit is found in records from several different parts of the world, because it was based on the length of a human forearm, which was much the same wherever the human happened to live. There is no reason to suppose that there was such a thing as a standard cubit. In practice each group of workers would have their own cubit-measure, probably made of wood, and probably determined originally by the forearm of a local person of importance. For large building projects, such as the Egyptian pyramids, the cubit-measure must have been officially determined, and its size carefully maintained throughout the entire operation. A few objects known as 'royal cubits' have survived in Egypt, and they may have been used in this way.

These comments apply also to the unit we call a yard, which was based on a stride-length, and, like the cubit, may well have been used long before we have any written records of it.

Another conclusion that can be drawn from our current knowledge of early measures is that different units were used for different purposes. There was no one unit of length (or capacity), from which all other units of length (or capacity) were derived by multiplication or subdivision. A cuneiform tablet may tell us, implicitly or explicitly, that thirty fingers make one cubit, but that only expresses the relationship between these units at one time and place. Such statements were not official definitions, they were conventional relationships, useful for numerical computations. Indeed, we should not be surprised to find two different units used for the same purpose in the same place at the same time. Attempts to discover universal systems of measures are, at best, misleading.

It might be thought that the foregoing remarks are fairly innocuous, but some nineteenth-century antiquarians adopted a very different viewpoint, and their ideas are still current today, especially on internet sites where the truth is what you want it to be.[9] The claim was that units such as the cubit were based on some prehistoric wisdom, possibly delivered to humanity by divine revelation, and carefully preserved by the wise-ones of the ancient world. One of the leading exponents of this theory was Charles Piazzi Smyth. His book *Our Inheritance in the Great Pyramid* (Figure 2.13) put forward the theory that the dimensions of this iconic building revealed knowledge of a standard unit he called a 'pyramid inch'. Not only was this unit derived mathematically from some mysterious knowledge of the size of the Earth, it was (he claimed) almost exactly the size of the standard British Imperial inch, as it was defined in the nineteenth century.

Piazzi Smyth's dreams were shattered when William Flinders Petrie re-measured the Great Pyramid and obtained numbers that could not be reconciled with the pyramid inch. Petrie later became famous for introducing systematic methods of archaeological excavation, and he wrote extensively on weights and measures. Unfortunately some of his own conclusions were based on flights of metrological fancy, such as the following.

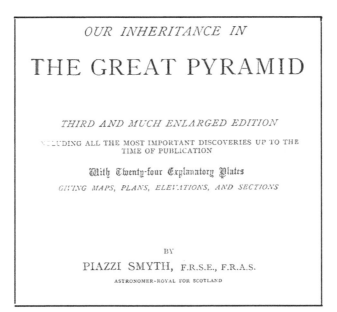

OUR INHERITANCE IN

THE GREAT PYRAMID

THIRD AND MUCH ENLARGED EDITION

INCLUDING ALL THE MOST IMPORTANT DISCOVERIES UP TO THE
TIME OF PUBLICATION

With Twenty-four Explanatory Plates

GIVING MAPS, PLANS, ELEVATIONS, AND SECTIONS

BY

PIAZZI SMYTH, F.R.S.E., F.R.A.S.

ASTRONOMER-ROYAL FOR SCOTLAND

Figure 2.13 The title page of Piazzi Smyth's book, 1877.

The largest family of lineal measures is that of the royal cubit of Egypt, 20.62 inches (524 mm). The double remen or diagonal of the cubit 29.161 inches (740.69) is almost exactly the natural length of a pendulum that swings 100,000 times in the day: at the latitude of Memphis this would be 29.157 (740.57). The close relation of 29.161 and 29.157 seems very unlikely to be a coincidence.

As we have seen, units of measurement appear to have had more humble origins. They were based on local practices, and varied in time as well as place. As civilization took a firmer hold, the rulers of empires and kingdoms would try to assert their power by making laws about the units, and issuing standard objects to define them. The royal cubits of Egypt may be examples of this practice, but they were surely based on an arbitrarily chosen length, such as the Pharoah's forearm, rather than the theory of the oscillation of the pendulum.

In Mesopotamia the need for standardization of weights and measures had been recognized by about 2100 BC. From this period there are tablets inscribed with hymns to the goddess Nanshe, giving thanks for the standardization of the size of the reed basket (possibly the sila measure mentioned above), and a measure known as a *ban*, believed to be 10 silas. Nanshe was the deity responsible for social justice, which suggests that uniformity of these measures was seen as a means of ensuring fairness in the distribution of grain. The hymns to Nanshe also refer to standard weights for weighing silver, the reason for which is rather more complex, as discussed later in *The elements of finance*. These Old Babylonian standards had a long legacy. In the British Museum there is a weight-object[10] made about 1500 years later. The inscription tells us that it was the property of Marduk-shah-ilani, and it is a copy of a weight issued by Nebuchadnezzar II, king of Babylon 604–561 BC. Significantly, the inscription also asserts that the weight is in accordance with the standard established by Dungi, who was king of Ur from 2096 to 2048 BC. We can conclude from similar evidence, both documentary and artefactual, that in some parts of the world the practice of weighing had advanced well beyond the primitive non-numerical stage by the start of the second millennium BC. Weighing in numerical terms requires combinations of weight-objects, so that a number can be assigned to any given amount of material. At first a set of equal stones could be used, so that two stones of material, or three stones, or however much is needed, could be measured out. In due course, by using stones of different sizes, it would become possible to measure a range of weights more easily and more accurately.

Before leaving the subject of standardization it is worth pointing out that attempts to enforce the uniformity of weights and measures have traditionally met with limited success. The First Emperor of China, Qin Shihuandi (221–210 BC), asserted his power over the newly united states by making proclamations requiring uniformity of weights and measures. His successor issued standard weights inscribed with edicts from both emperors calling for compliance with the official standards. However, we should retain a healthy scepticism about the effectiveness of these attempts at standardization, even under the rule of a ruthless warlord. In modern times, with better communications, a compliant people, and a higher level of literacy, the difficulties remained. Thus in 1813 the British reader of Cullyer's *Gentleman's and Farmer's Assistant* (Figure 2.9) would have been comforted by the fact that the unit of length, the yard, was indeed standard throughout the kingdom. But the Gentleman and Farmer would have been less happy to find that there were great variations in the size of the bushel-measures used for grain. In fact Cullyer felt it necessary to include a lengthy table by which his readers could determine how much 'the Bushel in Common Use contains more or less . . . than the Standard Winchester Bushel'.

The elements of finance

Our current ideas about the origins of money are based partly on the practices of the primitive tribes that still exist in the modern era, and that evidence is not necessarily conclusive. Fortunately, more reliable evidence can be found in the written records of the Old Babylonian period. For example, a clay tablet dating from around 1950 BC is inscribed with a proclamation from the king of Eshnunna. It provides explicit evidence of the various functions of money at that time. The primitive social functions are represented by fines for causing injury. For example, if you were guilty of 'biting a man's nose' you would have to pay about 500 grams of silver, but a 'slap in the face' would cost you less. There are also examples of the commercial functions of money, in the form of the recommended prices in silver for measured amounts of common goods.

It is clear from this tablet, and many others, that the use of silver as a form of money was well established by the beginning of the second millennium BC. Like grain and sheep, silver had an intrinsic value but, unlike grain and sheep, it was easy to transport and keep secure. Evidence of silver mining as long ago as the fourth millennium BC has been found in Turkey, and clearly significant quantities of silver had flowed into Mesopotamia. Originally that would have been a natural result of the fact that high-ranking people were addicted to what we now call bling, but even though the intrinsic value of silver stemmed from human vanity, its desirability was also what made it suitable for more basic, monetary, purposes. When silver and other precious metals are used in this way, we call them *bullion*. Of course, most people in Mesopotamia did not have any silver, and for them alternative forms of money were used. The Eshnunna proclamation states that the daily wage for a harvest worker should be a specific amount of either silver or grain. The unit of silver-weight was what we translate as a *shekel*, about 8 grams in modern terms, rather less than the weight of a current British one-pound coin (which is not made of silver, of course). A document from about 2000 BC, purporting to be based on correspondence between the king of Ur and his agent, tells us that the exchange rate was one shekel of silver to one gur of grain, where a gur is 300 sila.[11]

As we have seen, bullion was a rather special form of money, even in the advanced civilizations of the ancient Near East. In some other parts of the world the amount of precious metal available was very small, and alternative forms of money must have been used. However, in Egypt there was a plentiful supply of the most desirable metal, gold, and finds dated to the fourteenth century BC contain both gold and silver.[12] Many of the items are small lumps of metal, and the obvious implication is that a piece of the appropriate size would be picked out and offered in payment for some commodity or service.

The relatively high value of gold and silver meant that the size of a payment had to be measured exactly, and that was often done by weighing. Sometimes a piece of gold or silver wire would have to be cut to make a payment of the required weight. The illustration (Figure 2.14) shows rings of precious metal being weighed on an equal-arm balance.[13] The weight-object is in the form of a bull, and was possibly made of bronze. The use of bronze would be consistent with the impression that this is not an everyday scene; rather it depicts an official procedure connected with the functions of money in the economic organization of society. Here, and wherever bullion played an important part in the economy, there would have been a strong incentive for measurement by weighing, and the associated arithmetical procedures.

Clearly there were many opportunities for persons competent in arithmetic. Not only did weighing require a high degree of precision, the associated calculations had to be done with similar accuracy, to avoid sharp practice. Such high levels of precision were not required for measuring commodities like grain, but here the problem was that so many different units were in use. For grain, we have already mentioned the volume-sar, the sila, the ban, and the gur, each of which was convenient for a specific part of the economic process, but which had to be

Figure 2.14 Weighing money in ancient Egypt.

converted by arithmetical rules so that government and trade could operate efficiently. Conversion between different commodities also led inevitably to arithmetical problems. For almost the entire period of recorded history the teaching of elementary mathematics has focused on problems like these.

Increasing economic complexity also led to the development of secondary financial procedures. In Mesopotamia we read of loans of silver on which interest was to be paid in grain. The rate of interest was obviously a matter of some concern, and here too the associated calculations had to be done properly. It may be misleading to speak of 'banks' in the second millennium BC, but there can be no doubt that there were institutions carrying out some of the functions of modern banking houses. Arithmetic inevitably played its part in their operations, and we shall have much more to say about this kind of 'financial mathematics' in later chapters.

CHAPTER 3

From Tax and Trade to Theorems

On some journeys you reach a point where certainty and doubt collide. As a mathematical time-traveller this would have happened to you in the age of the classical Greek civilization, around 500–200 BC. You would have been certain of the truth of elementary arithmetic, but in geometry there might be some doubt. How could you be sure that the sum of the angles in a triangle is always equal to two right angles? So, with some diffidence, you might have set out to answer apparently simple problems, such as calculating the length of the diagonal of a square. And if you persisted, you would have discovered something startling.

The coming of coins

In Chapter 2 we saw how pieces of precious metal were used as money in the second millennium BC. In the places where this so-called *bullion economy* existed, the pieces of metal had to be valued by weighing them and, in many cases, doing some calculations. Both these operations, weighing and calculating, were tiresome, so we can be fairly sure that the bullion economy was not a feature of everyday life for most people at that time.

Throughout the next thousand years or so the world changed in many ways. Cities and kingdoms came and went, and trade of all kinds increased enormously. Archaeologists have scraped away at only a few layers of the evidence, and significant new discoveries may well alter our picture. However, there are already some indications of alternatives to the bullion economy. For example, there is evidence that cowrie shells (see Figure 1.3) were being used as money in China before 1000 BC, and for several centuries 'cowrie' was synonymous with 'money' in Chinese.

We refer to cowries as *token* money because they have little intrinsic value. They are easily portable, and as they are all practically identical, there is no need to weigh them. So counting can replace calculation for multiple purchases: one loaf for one cowrie means five loaves for five cowries, and the primitive notion of a one-to-one correspondence is sufficient for such transactions. When these features of cowries are imposed on lumps of metal, we refer to the lumps as *coins*. In other words, coins have uniform sizes, and can be used in trade without the need for weighing and complex calculations. This is an idea that evolved gradually over centuries.[1] From the ancient Near East there are a few examples of lumps of silver that have been marked in certain ways, presumably to indicate their value. But more significant steps towards coinage as we know it were made in the kingdom of Lydia (now in western Turkey) around 650 BC. The Lydians used small lumps of precious metal that have some, but not all, of the characteristics of coins, the main feature being that they have been marked in various ways. As the objects differ in size, it could be that the marks indicate different weights, but if so the system is unclear. The economist Adam Smith, in his *Wealth of Nations* (1776), made another suggestion. The Lydian objects are made of electrum, a naturally occurring alloy of gold and silver. As gold is much more valuable than silver, the relative proportion of the two metals affects the value of a lump of electrum, and Smith conjectured that the marks are related to this property, rather than the weight. The determination of the alloy in a coin is a complicated business, and we shall have a lot more to say about it in due course. Around the same time, different kinds of coins began to appear in China. They were made of bronze, which was regarded as a precious metal in that region. Some of the earliest ones were shaped like cowries, but there were also some strange shapes, including spades and knives.

The Chinese coins were a separate development, but the lumps of gold-silver alloy made in Lydia were soon followed by improved versions, produced in several parts of Greece. These were true coins: they were made of good silver, of uniform size and weight, and stamped on both sides with marks that identified their origin. In due course such coins became symbols of the importance of the issuing authority, and highly skilled artists were employed in their design and production. Coins like the one shown in Figure 3.1 made an impact for their aesthetic quality, as well as their economic usefulness, and similarly impressive coins were soon being produced elsewhere, in Persia for example. By the middle of the first millennium BC, the social and economic conditions in several parts of the world were ready for

Figure 3.1 An Athenian silver coin, fourth century BC. This coin is a silver tetradrachm.
Wikimedia Commons: supplied by Classical Numismatic Group, Inc., http://www.cngcoins.com.

the establishment of a system of coinage, more or less as we know it now, but with some notable differences. This development was to have important consequences for mathematics.

Practical mathematics

The development of coinage was just one of the innovations that were made in the Greek world from about 600 BC onwards. That world encompassed most of the lands bordering the eastern end of the Mediterranean Sea, including modern Turkey and parts of Egypt, and it was here that a new vision of mathematics began to evolve. However, it is a mistake to think that the new mathematics entirely displaced the old. The ancient wisdom that the Greeks had acquired from both Egypt and Mesopotamia played an essential part in creating their world, from its centre, the magnificent buildings in Athens, to the edges of the vast empire of Alexander the Great. That knowledge was eventually passed on to the Romans, who built another vast empire, renowned for the efficiency of its technology and administration. So before we begin to tell the story of the new mathematics, it is worth spending a little time describing how the old mathematics developed in the Greek and Roman periods.[2]

One of the main advantages of coins was that they facilitated the collection of taxes. The principle of taxation had advanced beyond the prehistoric form of 'protection money' to a form of social contract, where citizens and landowners made payments to a central authority in return for various benefits. When taxes were paid 'in kind', with agricultural produce, supply and demand were not always in balance, and there were obvious logistic problems. The use of bullion made for a more efficient process but, as we have noted, the precise valuation of lumps of precious metal was difficult. If the central authority could insist on taxes being paid in coins that the state itself had provided, then they would be sure of receiving the right amount.

The drawing in Figure 3.2 is one of the scenes depicted on the famous 'Darius vase'.[3] It represents tax being collected in a part of the Greek world, and it provides some tantalizing glimpses of practical mathematics at that time. The collector holds in his left hand an object that looks rather like a modern computer: indeed, it may actually be an ancient precursor, containing a list of the taxpayers and what they owed. On the table the collector appears to be writing some Greek letters. Some artistic licence has been invoked here, but these letters are surely intended to stand for numbers. This is a new way of representing numbers, and it is based on a new way of writing: a phonetic alphabet. Instead of using signs for objects and actions, the signs now represent sounds. The idea of an alphabet was not invented by the Greeks, but they had adopted it by the eighth century BC, and using the letters of the alphabet to represent numbers does appear to have been a Greek invention.[4] Remarkably it seems to have happened at the same time (around 650 BC) and in the same region (Asia Minor) as the invention of coinage.

Figure 3.2 Sketch of a scene on a Greek vase, c.330 BC.

Following the Egyptians, the Greeks used a system of numeration based on the number 10, which we call a *decimal* system. The numbers from 1 to 9 were represented by the first nine letters of the alphabet (Figure 3.3). The numbers 10, 20, 30, . . . , 90 were represented by the next nine letters, and the numbers 100, 200, 300, . . . , 900 by the last nine. (Because 27 letters were needed, and the standard Greek alphabet had only 24, three obsolescent ones were added.) With this notation, numbers up to 999 could be represented by writing the appropriate letters in order; larger numbers were indicated by special marks. This system was, in some ways, an improvement on the Mesopotamian and Egyptian methods of writing numbers. But it must be stressed that it had no particular advantages when it came to solving problems involving addition, subtraction, multiplication, and division. For example, what is twice ΩΜΓ?

So how did the Greeks actually do arithmetic? Hard evidence is surprisingly difficult to find. A Greek 'multiplication table', using the alphabetic numerals has survived, but it is an isolated example, and we have no documents referring to the use of such tables. The reasonable conclusion is that most arithmetic was done with a very primitive computing device, a set of pebbles. Using this method the operations of addition and subtraction can be carried out in ways that are almost self-evident; so much so that children are often taught in this way. Multiplications

	1	2	3	4	5	6	7	8	9
	A	B	Γ	Δ	E	F	Z	H	Θ
	α	β	γ	δ	ε	F	ζ	η	θ

Figure 3.3 The Greek letters (upper and lower case) for the numbers 1–9.

60 How many columns?

Figure 3.4 Finding the result when 60 is divided by 5.

and simple division are also quite straightforward. For example, suppose we want to divide sixty loaves equally among five groups of workers. We count out sixty pebbles, and place them in columns of five, as in Figure 3.4; then we count the columns, giving the answer, twelve.

Pebble-arithmetic is so easy and effective that it must have been familiar to those engaged in arithmetic back in the days of the early civilizations, or even before that. That there is no artefactual evidence from that era is hardly surprising, because the pebbles that were used would not be easily identifiable by modern archaeologists. But there is some evidence that by the time of the Greeks the pebbles were sometimes laid out on a counting-board, which was marked in columns to facilitate the procedures. A few of these counting-boards have survived, but they are rare.[5] There are also a few passing references to pebble-arithmetic in Greek literature, for example in the play *Agamemnon*, written by Aeschylus in the early fifth century BC.

Despite its very basic character, it is likely that pebble-arithmetic was the catalyst for a new kind of mathematical activity: the study of numbers for their own sake. For example, the Greeks were interested in *triangular* numbers (Figure 3.5). These are the numbers 3, 6, 10, 15, and so on, that can be made into neat triangular patterns when represented by pebbles.

Similar patterns, such as those representing the *square* numbers 4, 9, 16, 25, and so on, suggested links with the Greek studies of geometry, which we shall come to shortly. Another fruitful idea arose when the pebble-arithmeticians noticed that, while numbers like 8, 9, 10, and 12 can be conveniently represented by neat patterns of various kinds, some awkward numbers like 11 cannot (Figure 3.6). Here we see the shoots of new mathematics growing out of the roots of the old.

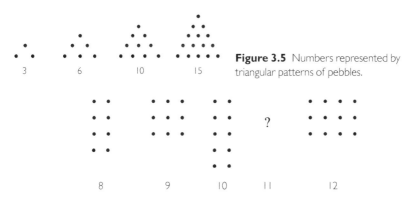

3   6   10   15

Figure 3.5 Numbers represented by triangular patterns of pebbles.

8   9   10   11   12

Figure 3.6 The number 11 cannot be represented neatly.

Although there were many political upheavals in the latter part of the first millennium BC, the cultural heritage of the Greeks was passed on to the Romans with little change. The Latin language gradually replaced the Greek one, but the two languages and the two alphabets co-existed for many centuries. From the mathematical point of view, the most significant change was the replacement of the Greek numerals by the Roman numerals. The basis of the Roman system was different, because there were far fewer characters than the 27 Greek ones, and numbers were represented by forming combinations of them in various ways. Nowadays we still recognize the sequence:

I II III IV V VI VII VIII IX X

because these symbols remain in common use, for purposes such as indicating the hours on clocks. However, like the Greek numerals, the Roman ones were mainly intended for recording information, rather than carrying out arithmetical operations. For practical purposes, the Romans continued to use pebbles (counters) and a counting board. The Latin word for a pebble was *calculus*, and the Latin word for a counting-board was *abacus*. The first of these has given us our word 'calculation', but the second creates problems for us, because it has been applied to other counting mechanisms. Many people now automatically think of an abacus as a frame on which beads slide along wires. That mechanism was not invented until the late Middle Ages, in China. For the avoidance of doubt, subsequent uses of the word 'abacus' in this book must be understood as referring to a board or cloth on which counters can be laid out.

The vast expanse of the Roman Empire provided many opportunities for the use of arithmetic. In northern Britain, on the boundary of the empire, the Romans built Hadrian's Wall, a defensive structure where a large garrison was stationed. In the 1970s archaeologists working there at a fort called Vindolanda found large numbers of wooden tablets, which have been dated to around 100 AD. They had been used for writing personal letters and keeping records of everyday events. The tablets as found were illegible, but scientific techniques eventually revealed an enormous amount of detailed information about the daily lives of the community.[6] Here is an example, part of Tablet 343.

> Octavius to his brother Candidus, greetings.
>   The hundred pounds of meat from Marius, I will settle up. From the time when you wrote to me about this matter, he has not mentioned it to me.
>   I have several times written to you that I have bought five thousand m[odii] of ears of grain, on account of which I need cash [. . .], at least five hundred [denarii].
>   See with Tertius about the 8½ [denarii] which he received from Fatalis.

This message contains several items that provide insight into the custom and practice of measurement and money, and the associated arithmetic, at that time. First we have the matter of 'a hundred pounds of meat', where the amount is written in full, in words (*pondo centum*). Then there is the matter of 'five thousand m[odii] of grain', where there is some abbreviation (*m quinque milia*). As might be expected, meat was measured by weight (the Roman pound of about 324 grams), whereas grain was measured by volume (the modius of about 8.7 litres). When sums of money are mentioned, the unit, the silver *denarius*, is understood: for example, Octavius's request for 'at least five hundred [denarii]' appears as *minime quingen-tos*. This seems to be the convention, whether the denarius is used as a measure of value or as a medium of exchange, so that a debt of 8½ denarii appears as *viii s*, the *s* standing for one-half.

The military might of the Romans allowed them to impose on their subjects whatever kind of monetary economy suited their purposes. Originally the den-arius was minted from good silver, and its weight was carefully controlled, but by about 250 AD the coins were being debased by adding significant amounts of cop-per. Eventually these coins became tokens, like cowrie shells, and their weight was irrelevant. The people had little choice but to accept these coins, and it seems that life continued quite satisfactorily, because large numbers of debased denarii are found throughout the Roman Empire. A hoard found near Frome in the south of England in 2010 contained over 50000 of them.[7]

The Vindolanda tablets provide a fascinating insight into life at the height of the Roman domination of the Western world, and the size of the Frome hoard suggests that arithmetic continued to play an important part in the government of the people. But it must always be borne in mind that the Romans did not use our modern notation for numbers, or our methods of doing arithmetic. The origins of these modern methods will be described in Chapter 4, but first we must turn back to a quite different form of mathematical activity.

The new mathematics

What is 27 plus 46? It has always been taken for granted that there is only one right answer to a question like this; but we can all make mistakes, so it is wise to check. One common method of checking is to do the calculation another way. If we have used our rules for addition and found that 27 plus 46 is 73, then we could apply our rules for subtraction and verify that 73 minus 27 is 46. In the Rhind Papyrus a numerical answer is often followed by a sign for 'do it thus', which indicates a check of that kind. This rather weak kind of proof was still being used in that sense in printed books in the nineteenth century.

The outstanding achievement of the Greeks was that they thought more deeply about the nature of proof. It is ironic that, although their studies began with philo-sophical speculations, the outcome was that mathematics became elevated beyond the level of a technique for working out tax returns and paying wages. It became the master key for the study of all human activity, whether it be travelling to the moon or hitting a golf ball.

The problem for the modern historian lies not in assessing the product, the achievements of the Greek mathematicians; rather it concerns the process, understanding the origins of those achievements and putting them in a firm context. There is, quite literally, a lot of de-mythologizing to be done. For example, many readers of this book will have heard that someone called Pythagoras proved an important theorem about right-angled triangles. Not so. The status of Pythagoras as an historical figure is dubious, to say the least, and the theorem was not 'proved' by anyone at the time (sixth century BC) when, it is claimed, he lived.[8] However, it is quite possible that the result was known to those who used geometry as a tool in practical work. The same can be said of some other assertions that used to figure prominently in popular accounts of early Greek mathematics. Happily, we can be fairly certain that someone called Euclid was living in Alexandria and writing around 300–250 BC, although that is essentially all we know about him (even the gender is a conjecture). There are near-contemporary references to Euclid's great work, known as the *Elements*, but that is essentially all we have.[9] Most of our knowledge of its origins stems from the account given by Proclus, who lived in Athens over 700 years later, and for the text itself we have only translations of copies of translations of copies. . . . The upshot of all this is that when we use the word 'Euclid' we really mean a book, not a person. And we must remember that, although the book was first compiled around 250 BC, what we refer to is a version based on much later translations and embellishments.

For the reasons given above, the following does not pretend to be a verbatim account of what Euclid actually wrote. At the start, the *Elements* lays the foundations for the study of geometry, by listing three kinds of statements. First, there are Definitions of geometrical objects, such as a point, a straight line, and a right angle. Then there are five Postulates describing what is assumed about these objects; the first one asserts that it is possible to draw a straight line joining any point to any other point. Finally, there are some Common Notions. These are statements that can be taken for granted as logical facts, such as the statement that two things which are both equal to the same thing are also equal to one another.

From this foundation, Euclid derives the whole of elementary geometry. The method is to prove propositions (theorems), each of which can be traced back to the postulates through a chain of similar propositions. A good example is the proposition that when two straight lines AB and CD intersect at a point X, the angles AXC and DXB are equal (Figure 3.7). This result is proved by appealing

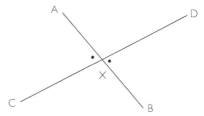

Figure 3.7 The marked angles are equal.

to an earlier proposition, which says that when one straight line cuts another, the sum of the two adjacent angles is equal to two right angles. So:

angle AXD plus angle DXB equals two right angles, because AB is cut by CD, and angle AXC plus angle AXD equals two right angles, because CD is cut by AB.

Now we can appeal to the common notions about equality, and what happens when we add and subtract equal quantities. Hence we obtain the result that the angles AXC and DXB are equal.

Some readers may object that the proposition proved above is obvious. But Euclid leads us far beyond the comfortably obvious into territory occupied by significant results that are by no means obvious. The so-called 'Theorem of Pythagoras' is one such result. Suppose we are given a triangle in which one angle is a right angle, and squares are drawn on its three sides, as in Figure 3.8. The theorem states that the area of the square on the side opposite the right angle is equal to the sum of the areas of the other two squares.

We do not know when this result was discovered, but there is clear evidence that it was widely known before it was proved by the methods found in Euclid. The proof depends on several chains of logical deduction winding back through many of the earlier propositions to the definitions, postulates, and common notions.[10] However, shining out from under this veil there is a mathematical gem, which can be appreciated for its own sake. The first step is one we have already mentioned: the fact that the area of a triangle is half that of a rectangle with the same base and height. This implies that, in Figure 3.9, the shaded square (1) is twice the size of the shaded triangle (2). Now comes the key move: turn this triangle through a right angle, keeping the corner (•) fixed. The result is another triangle (3).

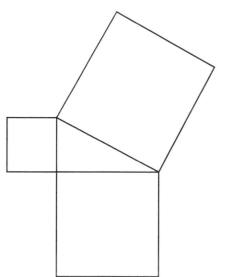

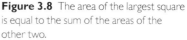

Figure 3.8 The area of the largest square is equal to the sum of the areas of the other two.

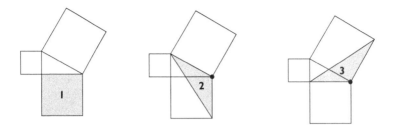

Figure 3.9 The first part of the proof of the Pythagorean Theorem.

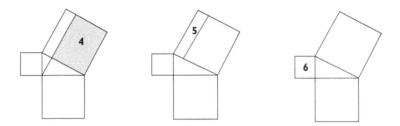

Figure 3.10 The second part of the proof of the Pythagorean Theorem.

This triangle is half the size of the rectangle (4) in Figure 3.10, as it has the same base and height. So the rectangle (4) is actually the same size as the square (1). As this rectangle is one part of the large square, we have only to show that the other part, the rectangle (5) is the same size as the square (6). This can be done 'similarly', which means that we can repeat the same argument, but with a different pair of triangles.[11]

Problems of measurement

'Euclid' is in fact a series of thirteen 'books'. The first four are devoted to the pure geometry of points, lines, and circles, using methods like those described above. The next topic is the relationship between geometry and arithmetic. Specifically, what does it mean when we say that a straight line joining two given points has a certain length? In the third and second millennia BC the answer to this question was that the length is defined in terms of a standard unit. It was accepted that the length may not be a whole number of units, so the unit must be divided into halves, thirds, quarters, and so on; indeed, some of these sub-units might have special names. This approach persisted into the modern era: for example, in the old British system a yard was divided into 3 feet, a foot into 12 inches, and so on. Associated with the idea of division into sub-units there was the primitive notion of a 'fraction', a device that is needed to make accurate measurements. But, like their Mesopotamian and Egyptian predecessors, the Greeks did not think of fractions as numbers in quite the way we do nowadays.

Unfortunately, there is a problem with this approach to measurement, and the Greeks discovered it. The simplest instance of the problem arises when we consider how to measure the diagonal of a square. Let us imagine how an intelligent Greek practitioner of mathematics might have looked at this problem in the fifth century BC. We shall call him Praktos. For the sake of exposition we shall use modern notation for numbers, although we must remember that Praktos would have had to struggle with the old Greek notation and methods of arithmetic.

Praktos might start his investigation by constructing an example, like Figure 3.11 where the square has side-length 100 units. How many units in the diagonal? According to 'Pythagoras' it should be the number whose square is the sum of the squares on the sides, that is $(100 \times 100) + (100 \times 100)$, which is 20000. As a practical man, Praktos might simply use his measuring stick, finding that the answer appears to be about 141, and he might well consider the possibility that the number is exactly 141, the discrepancy being because his square is not quite true. If so, he would have to resort to arithmetic, and work out 141×141, which turns out to be 19881, not quite large enough. He could then work out that 142×142 is 20164, which is too big. At this stage Praktos would conclude that his initial choice of 100 units for the side of his square will not work. He might try another choice, say 1000 units, and after doing some more tiresome calculations he would find that 1414 is too small and 1415 is too big. A few more unsuccessful trials might well lead him to give up. But if he were imbued with the Greek spirit of logical enquiry, he would begin to ask some awkward questions.

In fact, on his very first attempt Praktos could have spotted the basic difficulty. When the side of the square is 100 units, he could have seen *without doing any calculation* that the diagonal cannot be 141 units. This is because 141 is an odd number, and it follows that its square is also an odd number, so it cannot be equal to the required number 20000, which is even. Indeed, whatever the length of the side, *the diagonal must be an even number*, because its square is twice the square of the side.

If he reaches this point, the thoughtful Praktos is on the verge of a startling discovery. He knows that the diagonal is divisible by two, so its square is divisible by four. Hence twice the square of the side is divisible by four, and the square of the side is divisible by two. It follows, by an argument that we have already used, that *the side must be an even number*. But this is absurd.

Why? Because if both the side and the diagonal are even numbers, then we can take half of them and find another pair of numbers that also solve the problem.

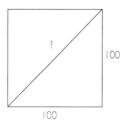

100

Figure 3.11 How many units in the diagonal?

For example, if it were true that a square with side 10000 has a diagonal of length 14142, then it must be true that a square with side 5000 has a diagonal of length 7071. But we have already ruled out this possibility, because one of these numbers is odd. Whatever the original numbers, it is clear than we cannot continue halving them indefinitely without reaching an odd number, and an impossible 'answer'.

What Praktos has discovered is that the diagonal and the side of a square cannot both be represented by whole numbers. This was known before Euclid's time, and was expressed by saying that the side and the diagonal of a square are *incommensurable*. It is the reason why four of Euclid's thirteen books are devoted to an exposition of lengths and the ratios of lengths, in an attempt to avoid the assumption that any geometrical length can be represented as the ratio of two whole numbers, or what we would call a 'fraction'.

When we try to summarize this part of Euclid, we must remember two things. First, we do not know exactly what Euclid wrote, or why; the version that has come down to us is rather like a school textbook, and contains few hints as to the motivation. Second, our summary is necessarily based on later insights, some of which were not discovered until long after Euclid. We should like the length of the diagonal of a square with side-length one to be a number, and we should like to call it the square root of two. But it cannot be a fraction, and if we are tempted to say that the square root of two is not a fraction, then we must explain what the square root of two really is. More generally, if there are numbers which are not fractions, how do we construct such numbers?

The diagonal of a square field was not the only awkward problem of measurement that the Greeks encountered. Another difficulty arose when they tried to measure the area enclosed within a circular field. When a field is bounded by straight lines the discovery of the rule for finding the area of a triangle removed the need for approximations, in theory if not in practice. However, the problem of measuring an area bounded by curved lines presented much greater difficulties, even in the apparently simplest case of a circle with radius equal to one unit.

The obvious starting point is to draw squares inside and outside the circle, as in Figure 3.12. The outer square has sides of length 2 units, and so its area is $2 \times 2 = 4$ square units. The area of the inner square is half that of the outer square. (This can be proved very simply, without calculation; if it is not clear to you, take a moment to think about it.) So the area of the inner square is 2, which suggests that the area of the circle is about 3. Indeed, this approximation can be found in several ancient texts, including the Bible. But even then it must have been clear, by direct

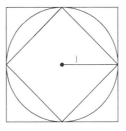

Figure 3.12 What is the area inside the circle?

measurement, that the true answer is slightly more than 3. Nowadays we express this conclusion by saying there is a number, which we denote by the Greek letter π, such that the area of the circle is π.

The mathematicians of Mesopotamia and Egypt did not use the π symbol, but they knew, in their own ways, that π lies between 3 and 4. When the Mesopotamians calculated the area of the circular base of a sila-measure (see Figure 2.11), their method was equivalent to taking π to be 3. The author of the Rhind Papyrus had a different method of calculating circular areas, which is equivalent to the assumption that π is $3\frac{13}{81}$. But the Greeks adopted a different approach, guided perhaps by the knowledge that there are some measurements that can never be expressed as fractions.

The Greeks regarded the determination of π as a geometrical problem. They asked if it is possible to construct a square equal in area to a given circle, using only the two simplest geometrical instruments, the ruler and compasses. In this context a ruler is simply a straight-edge; it has no markings like the modern school-ruler. Much of Euclid is devoted to constructions of this kind. For example, it is easy to solve problems like constructing a square with a line-segment of unit length as one of its sides, and hence the 'square root of 2' can be constructed as the length of a diagonal of this square. However, the Greeks were unable to solve the similar problem of 'squaring the circle'; that is, constructing a square with area the same as that of a circle.

Despite the failure to square the circle, the Hellenistic period did see some very significant progress on the problem of π. It was the work of Archimedes, the greatest mathematician of that period, who lived in the third century BC. He began by establishing that there is a relationship between two different problems about measuring circles. The first problem, as mentioned above, is to find the area of a circle of unit radius, and we have agreed to call this π. The second problem is to find the circumference of this circle. Nowadays every child knows (or should know) that the circumference is twice the same number π. But very few people of any age will be able to explain why the answers are related in this way. Archimedes proved it, by means of the following result: the area of a circle is equal to the area of a triangle with base equal to the radius of the circle, and height equal to the circumference. Given this result, the *half the base times the height* rule for the area of the triangle leads to our formula 2π for the circumference of a circle with unit radius.

Archimedes was also able to show that π lies between $3\frac{10}{71}$ and $3\frac{1}{7}$, a surprisingly accurate estimate, and way ahead of its time. It was not until 2000 years later that it was finally shown that the problem of squaring the circle is actually unsolvable. This does not mean that π does not exist; it means that it is not a number that can be constructed by the geometrical (ruler and compasses) methods used by the Greeks. So Archimedes' strategy of approximation is the best possible way of understanding what the number π really is.

Archimedes made several other amazing discoveries about areas and volumes bounded by curved lines and surfaces, but they were so far ahead of his time that we shall defer discussion of them to Chapter 6. He was also responsible for many significant contributions to the advancement of mathematics as a whole, and to the general understanding of the power of the subject. As an example we can cite his *Sand Reckoner*, a work believed to have been written around 220 BC. In this work, Archimedes discussed the problem of the number of grains of sand in the 'universe'. Some of his contemporaries thought that this number was so huge as to be beyond comprehension, but Archimedes explained how it can be described with the existing methods of representing numbers. He began by pointing out that everyone accepts that there are numbers, such as 1000000, which despite being large are nevertheless comprehensible. Given such a number we must therefore be able to comprehend the idea of carrying out an operation that number of times. In particular we can understand what it means to multiply that number of numbers. This holds even when the numbers themselves are large. So if we want to find the number of grains in some large object, we can simply fill a small pot of known size with sand, and count the number of grains in it. Then by repeated multiplication we can work out how many pots will fill up a larger pot, and an even larger pot, and so on. In this way we can estimate the number of grains that would fill our planet, and even the entire 'universe', as Archimedes understood it.[12]

Proving results about whole numbers

The ancient experts in pebble-arithmetic knew that some whole numbers can be divided into equal parts quite easily: for example, a set of 12 items can be split into 2 parts, 3 parts, 4 parts, or 6 parts without any items left over. The number 60 has even better properties in this respect, as it can be divided exactly into 2, 3, 4, 5, and 6 parts (and indeed into 12, 15, 20, and 30 parts as well). This is almost certainly the reason why the numbers 12 and 60 appear so frequently in ancient systems of measurement. On the other hand, some numbers are distinctly awkward when it comes to fair division (as illustrated in Figure 3.6). It is probable that the mathematicians of ancient Egypt and Mesopotamia noticed this, in the case of numbers like 7, 11, and 13, but the matter was not studied seriously until the time of the Greeks. The warnings given earlier in this chapter about attributing specific mathematical discoveries to specific 'Greek' mathematicians apply here too, but there can be no doubt that what we now call the Theory of Numbers began with the Greeks. In their terminology, the number 12 can be 'measured' by the numbers 2, 3, 4, and 6, meaning that a stick with any of these lengths can be used to determine the exact length of a line with length 12. We now say that 2, 3, 4, and 6 are *factors* of 12.

The earliest theoretical work on this subject is recorded in the 7th, 8th, and 9th books of Euclid's *Elements*, beginning with the observation that some numbers cannot be measured by any smaller number (except 1, of course). These are what we call the *prime* numbers. It is reasonable to exclude 1 itself, so the list of primes begins

$$2, 3, 5, 7, 11, 13, 17, 19, 23, 29, 31, 37, 41, 43, 47, \ldots.$$

Although the Greeks did not have our efficient notation for numbers, they were surely able to make a list like this. And so it is likely that some now forgotten mathematician in some part of the Hellenistic world was the first to suggest that:

the list goes on forever.

The truth of this statement is not obvious: if we continue the list as given above, we find that the primes begin to thin out quite noticeably. For example, as shown in Figure 3.13, there are 25 primes between 1 and 99, whereas there are only 14 primes between 901 and 999, a range of the same size. So the statement is open to either proof or refutation. The matter was settled (we must conclude) when some other forgotten mathematician in another part of the Hellenistic world came up with a marvellous proof.

This proof, as recorded in Euclid's book IX, is still regarded as one of the jewels in the mathematical crown. It goes as follows.[13] Suppose we could make a list of all the prime numbers: 2, 3, 5, . . . , P. Then we consider the number obtained by multiplying all these numbers and adding 1:

$$N = (2 \times 3 \times 5 \times \ldots \times P) + 1.$$

Adding 1 to the product is the master-stroke. It ensures that if we divide N by any member of our list there is always 1 left over, so N cannot be exactly divisible by any one of these numbers. Consequently there are just two possibilities. First, N itself could be a prime number, in which case we have constructed a new prime. On the other hand, if N is not a prime we can express it as the product of smaller prime factors. These prime factors cannot be in our original list, because that contains only numbers that do not divide N exactly, and so again we have found a new prime. In summary, we began with the assumption that we had made a list of all the primes, but we found another one, so our assumption cannot be true.

2	3	5	7					
11	13		17	19
	23			29
31			37	
41	43		47	
	53			59
61			67	
71	73			79
	83			89
			97	

		907	
911			919
			929
		937	
941		947	
	953		
		967	
971		977	
	983		
991		997	

Figure 3.13 The prime numbers between 1 and 99, and between 901 and 999.

There is another fundamental property of primes which, rather surprisingly, is not recorded explicitly in Euclid. As we have just noticed, any number that is not itself a prime can be expressed as a product of prime factors. The question is: could there be more than one way of doing it? For example, one way of breaking down 120 into primes is to start with 10×12:

$$120 = 10 \times 12 = (2 \times 5) \times (3 \times 4) = 5 \times 2 \times 3 \times 2 \times 2.$$

Another way is to start with 8×15:

$$120 = 8 \times 15 = (2 \times 4) \times (3 \times 5) = 2 \times 2 \times 2 \times 3 \times 5.$$

In both cases we get the same prime factors, but in a different order, which suggests that the factorization is essentially unique. However, the experimental evidence is not really conclusive. For example, the numbers 56909, 127643, 25657, and 283121 are primes, and

$$56909 \times 127643 = 7264035487, \qquad 25657 \times 283121 = 7264035497.$$

So these two products of primes are very nearly equal. It is not at all obvious that there could not be a similar example in which the products are actually equal. But in fact this cannot happen: every number can be expressed as a product of primes in only one way, apart from the order. This means that the primes are very special: not only is it possible to construct any whole number by multiplying primes, but it can be done in just one way.

There has been much speculation about why this result is not stated explicitly in Euclid. The extant version does contain some theorems from which the result follows quite easily, and it is possible that it was included in the original version, but was discarded as being of little importance by the later editors.[14] In fact, it seems that the Greeks and their followers regarded the primes as rather awkward, and preferred to study more likeable numbers, to which they gave approving names, such as 'perfect' or 'amicable'. Such numbers are quite interesting, but they are by no means as fundamental to mathematics as the primes. Strangely, it was nearly 2000 years before the theory of the primes was taken up again by mathematicians, with significant results, as we shall see in Chapter 7.

CHAPTER 4

The Age of Algorithms

Y ou have read how increasing levels of organization in society led to greater reliance on arithmetic and geometry. An inevitable consequence was the need to find better ways of dealing with numbers, and by the end of the first millennium AD improved methods had reached many parts of the civilized world. These methods were based on the notion of an 'algorithm'. Around the same time there emerged a new form of mathematical discourse, which we know as algebra. Together with the new ways of doing arithmetic, it formed the foundation for the mathematics that pervades our daily lives in the twenty-first century.

How to do arithmetic

Back in the third millennium BC there had been major advances in arithmetical procedures, particularly those used for the operation of division. But for over 2000 years after that there was little progress. The main function of arithmetic was to manage resources, by means of written accounts based on numbers and measurements. As the level of social structure increased, the accounts were used in more complex ways, designed to ensure that the resources were allocated and used efficiently. This early arithmetic had three components.[1]

The *notation* for numbers, including fractions;

The *procedure* for working out the answer. The procedures included addition, subtraction, multiplication, and division, as well as some more specialized operations;

The *medium* used. Several new media had superseded the prehistoric method of scratching marks on wood or bone. In addition to finger-reckoning and pebble-arithmetic, they included making wedge-shaped impressions on clay tablets, marking papyrus with a reed pen, and drawing lines in the sand.

Many different ways of combining the notation, the procedure, and the medium had been devised. But the historical record contains no hint of a system that could

Figure 4.1 Numbers 1–9 represented by Chinese counting rods.

be regarded as clearly superior to the others. In particular, the far-reaching achievements of the Roman era did not include a really good way of doing the computations on which those achievements were based. Undoubtedly, Roman scribes became adept at moving counters and reckoning on their fingers, but the lack of contemporary evidence of the detailed procedures suggests that theirs was an elite occupation. That arithmetic was a complex process helped to maintain the privileges of those who had mastered it.

The catalyst for change came from the east. By the fourth century AD the Chinese had developed a practical system of arithmetic based on counting-rods.[2] Each of the numbers from one to nine was represented by a simple configuration of rods, similar to those shown in Figure 4.1. The counting-rod configurations could also be used as numerals in written texts. For example, the Chinese character for five could be replaced by the symbol |||||, just as we use 5 instead of the word 'five'.

As well as being decimal (based on the number ten), the important feature of arithmetic with counting-rods was the use of a place-value system for representing large numbers. Unlike the Roman system, new symbols were not needed for ten, 100, 1000, and so on, because these numbers could be indicated by putting the basic numerals in the appropriate place (Figure 4.2).

For the purpose of calculation the numbers would be represented by placing the counting-rods on a board divided into columns, and arithmetical operations could then be performed by moving the rods in certain ways. In the case of addition, the procedures are almost self-explanatory: using modern symbols may help us to understand them, but the rods themselves display the meaning. For example, to add 325 and 67, rods are placed on the counting board in columns, representing (from right to left) units, tens, hundreds, and so on (Figure 4.3). First, the six in the bottom line of the tens-column are added to the two in the top line, making eight tens in all. Then the seven in the units-column are added to those in the top line, making twelve, so there is an overflow of ten, which results in adding one more rod to the tens-column and leaving two in the units-column.

By about 500 AD the decimal place-value system had reached India. We do not know precisely what procedures were used by the Hindus at that time, but it is

Figure 4.2 The numbers we now write as 325 and 67.

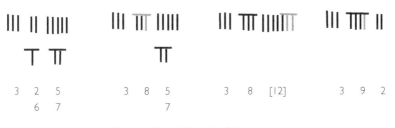

Figure 4.3 Addition with Chinese rods.

fairly clear that their practices differed from the Chinese in at least two significant ways: they used different symbols for the numbers, and their preferred medium for performing the arithmetical operations was the *dust-board*. This was a wooden tray covered with a thin layer of sand, in which the numerals were represented by making marks with a stick. In the course of a calculation, such as multiplication, a numeral would be recorded at one step, only to be erased later when that part of the calculation was complete. With the dust-board it was not easy to keep the columns in line, so the absence of a numeral had to be signified by a special symbol.[3] That may be the origin of our symbol 0 for 'none': 305 means three hundreds, no tens, and five units. At one time it was claimed that the invention of such a symbol was a major breakthrough, but it was not a new idea. A sign for nothing had been used in Old Babylonian arithmetic, and in the Chinese system it would have been indicated simply by not putting any rods or counters in the appropriate column of the board. More significant was that, in due course, the symbol 0 came to be regarded as a numeral, just like the signs for the numbers 1 to 9. With this device the Hindus now had a firm basis for efficient arithmetical procedures.

From the Old Babylonian period onwards it had been recognized that a difficult computation can be made easier if it is broken down into simpler ones, the results of which are already known and tabulated. The simplicity of the Chinese and Hindu systems meant that relatively few such tables were needed. Specifically, if the tables for adding and multiplying the numerals from 1 to 9 are known, then we can add and multiply any numbers, however large, by following some simple rules. These tables are small enough to be memorized. And, of course, knowing the tables is not enough: it is also necessary to apply the rules correctly.

An indication of the crucial importance of the Hindu system is the lack of contemporary advances in the Western world. In early medieval Europe, even in England, much effort was devoted to complicated calculations. The main topic was the calendar, particularly the date on which the Christian festival of Easter ought to be observed. However, there is no evidence of significant advances in the practice of such calculations.[4] Arithmetic was regarded with horror by contemporary English scholars, such as Aldhelm of Malmesbury, who wrote (*c*.670) of 'the near-despair of doing all that reckoning'. He was probably referring to the kind of finger-reckoning illustrated in a tenth-century manuscript copy of the Venerable Bede's *De Tempore Ratione*.[5] The contortions required to represent numbers larger

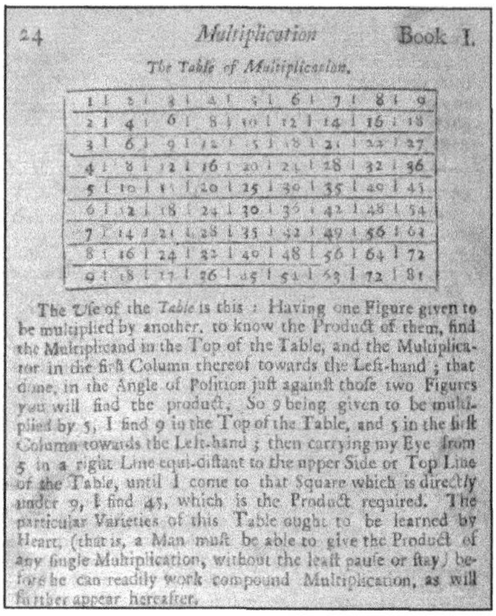

24 *Multiplication* Book I.

The Table of Multiplication.

1	2	3	4	5	6	7	8	9
2	4	6	8	10	12	14	16	18
3	6	9	12	15	18	21	24	27
4	8	12	16	20	24	28	32	36
5	10	15	20	25	30	35	40	45
6	12	18	24	30	36	42	48	54
7	14	21	28	35	42	49	56	63
8	16	24	32	40	48	56	64	72
9	18	27	36	45	54	63	72	81

The Uſe of the *Table* is this: Having one Figure given to be multiplied by another, to know the Product of them, find the Multiplicand in the Top of the Table, and the Multiplicator in the firſt Column thereof towards the Left-hand; that done, in the Angle of Poſition juſt againſt thoſe two Figures you will find the product. So 9 being given to be multiplied by 5, I find 9 in the Top of the Table, and 5 in the firſt Column towards the Left-hand; then carrying my Eye from 5 in a right Line equi-diſtant to the upper Side or Top Line of the Table, until I come to that Square which is directly under 9, I find 45, which is the Product required. The particular Varieties of this Table ought to be learned by Heart, (that is, a Man muſt be able to give the Product of any ſingle Multiplication, without the leaſt pauſe or ſtay) before he can readily work compound Multiplication, as will further appear hereafter.

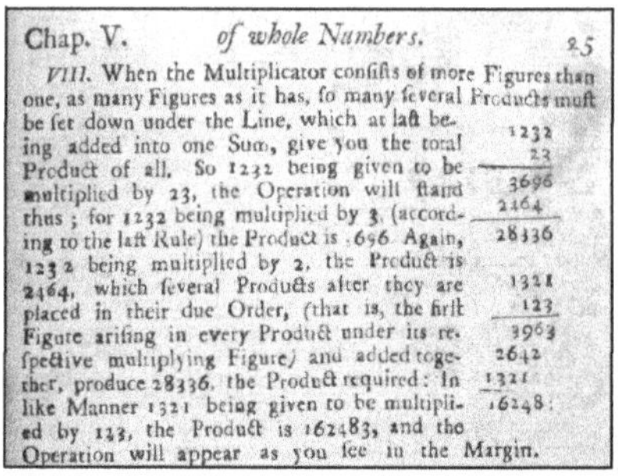

Chap. V. *of whole Numbers.* 25

VIII. When the Multiplicator conſiſts of more Figures than one, as many Figures as it has, ſo many ſeveral Products muſt be ſet down under the Line, which at laſt being added into one Sum, give you the total Product of all. So 1232 being given to be multiplied by 23, the Operation will ſtand thus; for 1232 being multiplied by 3, (according to the laſt Rule) the Product is 3696. Again, 1232 being multiplied by 2, the Product is 2464, which ſeveral Products after they are placed in their due Order, (that is, the firſt Figure ariſing in every Product under its reſpective multiplying Figure) and added together, produce 28336, the Product required: In like Manner 1321 being given to be multiplied by 123, the Product is 162483, and the Operation will appear as you ſee in the Margin.

```
 1232
   23
-----
 3696
2464
-----
28336

  1321
   123
------
  3963
 2642
1321
------
162483
```

Figure 4.4 'Long multiplication', described by Wingate in the seventeenth century AD.

than ten using finger-signs were grotesque, and this must surely have contributed to Aldhelm's despair.

Decimal place-value arithmetic was a long time in the making. It continued to evolve up to the seventeenth century AD, by which time it had taken on a form very similar to the current one. The typical example shown in Figure 4.4 is taken from a famous textbook[6] written by Edmund Wingate in 1630. Wingate's description of long multiplication (on the right-hand page) is a sequence of instructions, aimed at a student of average intelligence. If the instructions are followed faithfully, the answer will be correct, even if the student does not understand the underlying reasons.

For several hundred years people were taught to 'do arithmetic' in that way, but in the 1970s a new way became widely available: the electronic calculator. We know that this device makes it easier for people to perform calculations such as multiplication, but why? The answer is that the electronic calculator has been programmed with a sequence of instructions similar to Wingate's, so now we have only to enter the required numbers, and the calculator does the rest. By tracing the methods described by Wingate back to their origins, we can begin to evaluate the pros and cons of putting our trust in electronic devices.

Advances in arithmetic

Soon after 630 AD, large areas of northern Africa and western Asia came under Arab rule, in a remarkably short period of time. By the end of the seventh century this Islamic world included parts of India, where the system of decimal place-value arithmetic was in use. Consequently, descriptions of the Hindu system, written in Arabic, began to appear. Possibly the earliest, and surely the most famous, was the *Short Treatise on Hindu Reckoning*, written around 825 by a scholar whom we know as al-Khwārizmī. The characteristic feature of al-Khwārizmī's work was the careful description of step-by-step procedures for carrying out mathematical operations. For this reason his name is immortalized in our word *algorithm*. The word signifies an orderly sequence of instructions that can be given to any kind of computer, human or electronic, to produce the answer to a general problem, such as the multiplication of two large numbers. Together with the Greek concept of deductive proof, the idea of an algorithm was the foundation of all subsequent progress in mathematics. As we shall see, the relative importance of the two ideas varied with the fashions of the times, but in the twentieth century the importance of algorithms was renewed, with the introduction of electronic calculators and computers.

Historically speaking, we cannot rely totally on al-Khwārizmī's *Treatise*, because no Arabic version has survived. This has led to speculation about its true contents, including the suggestion that at least part was devoted to finger-reckoning. The earliest text on arithmetic of which we have a version in Arabic is probably one written around 950, in Damascus, by Abdul Hassan al-Uqlīdisī. His name is an Arabic form of Euclid, which reminds us that mathematics from Greece, as well as India, was being integrated into the Islamic tradition. As with al-Khwārizmī, there is no original copy of al-Uqlīdisī's text, but we do have a version in Arabic written about 200 years later. In this manuscript the algorithms for arithmetical operations, such as long multiplication, are described explicitly.[7] The Hindu-Arabic signs for numbers appear frequently, but not always in the form we use now. There is also what is thought to be the earliest known use of decimal fractions, a topic that we shall discuss in Chapter 6.

The manuscript also refers to a non-mathematical development of great practical significance: the use of a new medium, paper. We are told of the problems of using a dust-board, such as 'the exposure of the content to the blowing wind', and the recommendations instead to use 'inkpot and paper'. It is believed that the secret of making paper had been revealed to the Arabs by the Chinese. Although paper is similar to papyrus (made from reeds) and parchment (made from skins), it is much cheaper to produce. Its use in Islam is the main reason why we are on firmer ground, historically speaking, when we describe the mathematical achievements of that world.

Another early text in Arabic is the *Principles of Hindu Reckoning* written by a Persian scholar Kūshyā ibn Labbān, who was born around 975.[8] His book described in detail how to carry out the algorithms of arithmetic on a dust-board. For example, his method for multiplying the numbers 325 and 243 involved finding the products of the digits 3, 2, and 5, and the digits 2, 4, and 3, and adding the results in a certain way. On the dust-board it was easy to keep the products in their proper places, and the digits that were no longer needed could be erased as the calculation proceeded.

The procedure for 'long multiplication' described by ibn Labbān is based on the same principle as the method given Wingate's book 600 years later. The underlying

325 × 243	=	300 × 243	=	300 × 200	=	60000		
				+300 × 40	=	+2000		
				+300 × 3	=	+900	=	72900
		+						
							+	
		20 × 243	=	20 × 200	=	4000		
				+20 × 40	=	+800		
				+20 × 3	=	+60	=	4860
		+						
							+	
		5 × 243	=	5 × 200	=	1000		
				+5 × 40	=	+200		
				+5 × 3	=	+15	=	1215	=	78975

Figure 4.5 The principle of long multiplication, in modern notation.

working is set out in full in Figure 4.5, using modern notation. The central column shows the products of the digits, with their correct place-values indicated by the number of 0s. The different algorithms for long multiplication are just ways of organizing this calculation so that it becomes a fixed routine, applicable to any specific instance of the problem. Although we may never be able to retrace clearly the paths through which the algorithms of arithmetic were developed, the outlines of the trail, from China, through India and Islam, to Wingate, are still visible.

For future reference it is worth noting that, as the numbers 325 and 243 have three digits, the number of single-digit multiplications required in this example is $3 \times 3 = 9$. These nine steps determine the amount of time required to carry out the algorithm. If the numbers had four digits, there would be $4 \times 4 = 16$ steps, and so on.

The uses of arithmetic

The new algorithms spread gradually into Europe. The documentary evidence is patchy, and it may well be that some of the traditional accounts of this subject rely too much on a few extant documents that are not representative. Another factor that adds to the mystery is that those who knew how to do arithmetic often kept their knowledge secret. Documents and artefacts are still being discovered, and it is possible that the picture will change. Nevertheless, it is clear that the demand for new and more efficient methods of calculation came from the machinery of government and commerce.

During the initial period of rapid Islamic expansion the conquerors struck no coins of their own, but by about 670 they were producing coins imitating those previously used in the conquered territories. These coins ranged from Persian silver drachms to Byzantine gold solidi. However, it soon became clear that a more uniform coinage would help to ensure the economic stability of the empire. The revenue from taxation had to provide for a welfare system, specifically the payment of a state pension to people who had served in the army.

Figure 4.6 An Islamic gold dinar, ninth century AD. The coin is a dinar
of the caliph al Ma'mun. Wikimedia Commons: supplied by Classical
Numismatic Group Inc., http://www.cngcoins.com.

At the end of the seventh century the caliph 'Abd al-Malik introduced a new
Islamic coinage in gold and silver. The gold dinar (Figure 4.6) differed from earlier
coins in that it lacked any pictorial features, in accordance with the teachings of
Islam. Instead it had a long inscription in Arabic. Its mass was determined by an
Arabic standard of weight, about 4.25 grams in modern terms, slightly less than
the Byzantine solidus. The corresponding silver coin was known as a dirhem, and
weighed about 2.9 grams.

Both the actual weights and the ratio between them were closely controlled for
hundreds of years. Within the Islamic world the dinar and the dirhem were ideal
for the purposes of the government. The tax-collectors could insist on being paid
in the current coin, and so persons with old or foreign coins, or odd pieces of gold
and silver, would have to sell them to the government to acquire acceptable coins.
Clearly, this mechanism allowed the authorities to make a profit on the transactions, provided they had the arithmetical skill required.

On the boundaries of the Islamic world, trade with other nations was flourishing, and that presented different problems. Although coins were now commonly
used as money-objects, the coins issued by different kings and princes were many
and varied.

The value of a coin depended on the amount of precious metal that it contained,
and that in turn depended on two measurements, the weight and the *fineness*
(purity) of the metal. The measurement of weight was relatively simple, and coins
were routinely weighed in the course of trade. But the fineness of a coin was more
difficult to determine. Gold coins were a particular problem, because small variations in their fineness significantly affected their value. The traditional method of
checking the fineness of a gold coin was to make a small mark with it on a touch-
stone; the colour of the mark indicated the fineness of the gold. But this was not a
method that could be relied on in everyday commercial transactions, and in practice the valuation of coins was determined by a combination of trial, experience,
and tricky arithmetical calculations.

Similar problems had arisen in India, and several examples of this kind of
commercial arithmetic occur in a document known as the Bakshali Manuscript.[9]

Figure 4.7 What is the fineness of the gold obtained by combining these coins?

Although its exact date is uncertain, this document is a clear reminder of the arithmetical difficulties associated with trade in the early middle ages. It contains several relevant problems, of which the following is a very simple example. Suppose we have four coins, weighing 4, 3, 2, 1 units, and the percentage of gold in each one is as shown in Figure 4.7. If the coins are melted down together, what will be the percentage of gold in the resulting lump of metal?

Simple problems like this can be solved by trial-and-error methods, but in practice the problems would have been much more complicated. The mints where gold coins were produced must have employed skilled arithmeticians, who could be relied on to multiply and divide correctly. The numbers had to be specified exactly, and the calculations done with a high degree of accuracy.

The north-west frontier

On the north-western borders of the Islamic world the main form of money was silver. The debased Roman denarius had been minted in large numbers, and in some regions it may have remained in circulation for a long time after the Romans left, but when the native peoples of this region began to mint their own silver, they restored the ideal of a high standard of fineness.

An early tenth-century hoard of silver, found at Cuerdale in the north of England, provides some revealing insights into the economic practices of the time.[10] This was the era when the Vikings were in control of large parts of England, and the money was a combination of coins and bits of silver wire, rods, and broken ornaments. The exact purpose of the Cuerdale hoard is not known: it may have been intended to pay for the army and its provisions, or it may have been connected with the flourishing trade in slaves. The Vikings also resorted to the ancient practice of extorting large sums of money in return for leaving the natives in peace. This money was known as the Danegeld, and it must have required some arithmetical skill (on both sides) to administer it. Another area of finance that would have required arithmetical expertise of a high order was the eleventh-century system of minting the Anglo-Saxon silver coins, which involved periodic recoinages and changes of the weight-standard. The details of this system are still the subject of much debate among numismatists, but sadly the light that might have resulted from evidence of arithmetical practices is lacking. We know that the Vikings themselves came into contact with Islamic traders in eastern Europe, and indeed they adopted standards from that source. But their arithmetical methods remain a mystery, and we must look elsewhere for the first traces of Hindu-Arabic arithmetic in western Europe.

The earliest evidence comes from the fourth quarter of the tenth century, when the numerals began to appear in manuscripts written in Latin. A key figure is the French scholar Gerbert of Aurillac.[11] In the 960s Gerbert spent three years in Spain, and it is almost certain that he learned about the numerals and the associated algorithms at that time. Parts of Spain had been under the rule of Islam since the eighth century, and Hindu-Arabic arithmetic would have been well-known there.

In recent years new evidence has come to light, and we can now be fairly certain about the nature of Gerbert's innovations. The traditional Roman abacus was, it must be remembered, not the type with beads on wires; it was a board on which counters were placed in rows and columns. Gerbert incorporated features of Hindu-Arabic arithmetic in two ways. First, he created a new design for the counting board, using the decimal place-value system. Figure 4.8 shows the layout of part of Gerbert's abacus, according to a drawing made soon after the year 1100. There are groups of three columns, and at the head of each group of three columns one of the Hindu-Arabic numerals is written, but here they are used simply as markers and play no part in the arithmetic.

Gerbert's second innovation was to label the counters with the Hindu-Arabic numerals, so that the procedures of arithmetic could be carried out by moving the counters on the abacus. Both these innovations were simply improvements on existing methods, but they had a significant impact on western European arithmetic. It may well be that other scholars were involved, but we know about Gerbert, in particular, because he has another claim to fame in the annals of history. His reputation as a scholar spread widely, and he was employed by some of the ruling princes of his time. These connections eventually led to Gerbert being installed as Pope Sylvester II in 999. Sadly, as an outsider, his appointment was controversial; he was ejected from Rome in 1002, and died in 1003.

In the twelfth century beautiful representations of Gerbert's abacus began to appear in illuminated manuscripts, providing clear evidence that it was known to

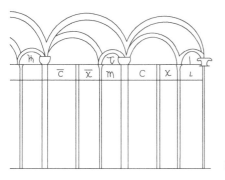

Figure 4.8 Layout of part of Gerbert's abacus.

the monastic orders.[12] It is reasonable to assume that it was used in the administration of their great estates, and it is tempting to suppose that it was also used at the higher levels of national government. Indeed, there is a great deal of circumstantial evidence to support this claim. Soon after Henry I became king of England in 1100, a major reform of the treasury at Winchester took place. The office for receiving taxes became known as the Exchequer, which has traditionally been thought to indicate the use of a counting-board resembling a chess-board. It is not clear how this differed from earlier counting-boards. The business of tax-collection consisted mainly of adding up large amounts of money, and that would require a counting-board specifically designed for dealing with the awkward monetary units of the time: pounds, shillings, and pence.[13]

But there are many ways in which Gerbert's more sophisticated abacus could have been used at the Exchequer, and at the Mints where coins were produced. For example, it would have been very useful for calculating the fineness of the metal from which the coins were made.

Although the details are vague, there can be little doubt that Hindu-Arabic arithmetic, using the medium of Gerbert's abacus, was well-established in northwest Europe in the twelfth century. In due course the abacus was to be superseded by the system of 'pen-reckoning', where the arithmetical procedures are performed by writing the numerals in prescribed ways on parchment or paper, rather than by moving numbered counters. That had been advocated by al-Uqlīdisī in 952, and it was the method described many centuries later by Wingate. In Chapter 5 we shall see how it played a part in the great flowering of art and learning that happened in Europe from the thirteenth century onwards.

The art of al-jabr

If the reputation of al-Khwārizmī rested only on his writings on Hindu arithmetic then it would be shaky, because there is no reliable evidence of what he actually wrote on that subject. But he has other, more secure, claims to fame. We know for sure that he wrote a book with a long Arabic title, including the word *al-jabr*, contemporary copies of which are extant.[14] The precise meaning of this word in its original context is unclear, but, eventually, *al-jabr* became *algebra*, the most powerful of all mathematical problem-solving methods. However, al-jabr did not begin to resemble modern algebra for many hundreds of years. So the reader who is nervous about elementary algebra and the mysterious 'unknown' x can read on with confidence; by following the historical threads, the significance of x will gradually become clear.

Al-Khwārizmī's book was possibly not the first to use the word al-jabr, and he did not say explicitly what it means. But he gives enough examples for us to understand the methods involved. The basic idea is that if we have two equal quantities, and we apply the same operations to both of them, then the final quantities are also equal. Conversely, if the final quantities are equal then so are the original ones.

More complicated forms of this kind of reasoning can be used to justify geometrical constructions, such as the ancient rule for finding the area of a triangle (see Figure 2.8). In that argument the triangle is duplicated and, after some cut-and-paste operations, the result is a rectangle that is plainly twice the size of the original triangle. In modern terms, al-Khwārizmī's method is equivalent to working with what we call 'equations', but that is not how he described it.

Literal translations of al-Khwārizmī's book on al-jabr must be treated with care, because we do not fully understand the historical context. Some of his terminology may not ring true when rendered into modern English. He refers to entities of three types, which can be roughly translated as follows.

Root: a quantity, such as a measurement, whose value is required
Square: the square of the root (so if the root is a length, the square is an area)
Number: a given numerical value

The problems are expressed in words as relationships between these roots, squares, and numbers. Here is one of his most famous problems, which we now call a quadratic equation.

Roots and squares equal to numbers: for instance 'one square, and ten roots of the same, amount to thirty-nine': that is to say what must be the square which, when increased by ten of its roots amounts to thirty-nine?

The solution is this: you halve the number of the roots, which in the present instance yields five. This you multiply by itself, the product is twenty-five. Add this to thirty-nine: the sum is sixty-four. Now take the root of this, which is eight, and subtract from it half the number of the roots, which is five: the remainder is three. This is the root of the square you sought for.

This method of calculation is not original; there are tablets from the Old Babylonian period that follow the same path, but without explanation or justification. Al-Khwārizmī makes it plain that the method is an algorithm, a sequence of instructions that leads from the given values, ten and thirty-nine, to the answer three. With any other given values, carrying out the same instructions will lead to the required answer. Indeed, it is worth pointing out that al-Khwārizmī's algorithm, when written in abbreviated form, is very similar to what we now know as a computer program.

INPUT: 10, 39

Half(10) → 5

Square(5) → 25

Add(39, 25) → 64

SquareRoot(64) → 8

Subtract(8, 5) → 3

OUTPUT: 3

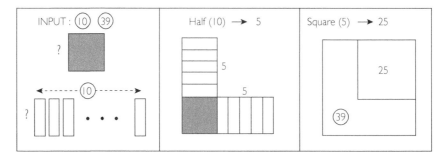

Figure 4.9 Completing the square.

How did al-Khwārizmī know that his algorithm was correct? The only method of proof available to him was classical geometry, and that was how he did it. The procedure begins by 'completing the square' (Figure 4.9). The first picture is the 'square and ten roots'. This can be thought of as a square plot of land, together with ten strips of unit width with length the same as the side of the square. First, 10 is halved to give 5, and two sets of 5 strips are placed as shown in the second picture. This produces a square with a piece missing, and we are told that the total amount of land is 39. The missing piece is a square of side 5, which amounts to 25. So the large square in the third picture has total area 39 plus 25.

The last part of the algorithm (Figure 4.10) depends on 39 + 25 = 64, which means that the side of the large square has length 8, the square root of 64. It follows that the side-length of the original square is obtained by subtracting 5 from 8, giving the answer 3.

Geometrical arguments like this require that the given numbers are positive, so al-Khwārizmī had to present separate rules for two other forms of quadratic equation. He referred to them as 'squares and numbers equal to roots', and 'roots and numbers equal to squares'. In all three cases the geometrical justification is based on the idea of completing the square. As with the algorithms of arithmetic, this beautiful application of geometry reached Europe by the many routes that

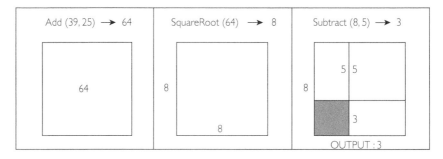

Figure 4.10 The conclusion of al-Khwārizmī's algorithm.

existed in the Islamic world. For example, a Hebrew text written by Abraham bar Hiyya (Savasorda), born in Barcelona in 1070, contains the solutions to all three cases of al-Khwārizmī's problem. It was translated into Latin by Plato of Tivoli around 1145.

The origins of symbolic algebra

In the old Islamic texts the procedures of al-jabr were described in words and justified by geometrical arguments, if they were justified at all. The algorithms can be elucidated by the use of modern algebraic symbols, but there are good reasons for not doing so at this point in the story. The symbolic method did not begin to appear until the thirteenth century, and it was not fully developed until the seventeenth, when it was deployed to great effect by Isaac Newton and others. Using the original forms of expression, we can see more clearly how the basic ideas developed and, incidentally, de-mystify the algebraic symbolism. On the trail from al-Khwārizmī to Newton we shall encounter many results that are now considered to be part of modern algebra, although they were discovered without its help.

One of the pioneers of the symbolic method was Jordanus de Nemore, the author of at least six mathematical texts. It is fairly certain that his work on arithmetic, *De Numeris Datis*, was written in the period 1220–1240, but we know very little else about him.[15] He seems to have been familiar with some Arabic texts, in particular ibn Labbān's *Principles of Hindu Reckoning* where algorithms were described by giving examples with specific numbers. Jordanus himself tried to give more general explanations, particularly when his questions were designed to teach the principles of arithmetic. For example, he considered the question of finding the parts when a given number is separated into two parts whose difference is given. For example, how can 10 loaves be split into two parts which differ by 2? He explained the method by the observation that 'subtracting the difference from the total what remains is twice the lesser'. Applying this method to the case when 10 is separated into two parts whose difference is 2, he subtracted 2 from 10, giving 8, and halved it, which gives the smaller part 4. Then, subtracting this from 10, the other part is 6.

Jordanus moved on to discuss a slightly more difficult question: if a given number is separated into two parts such that the *product* of the parts is also given, how can the parts be found? Here he departed from the traditional method by using symbols to represent the given numbers, without assigning specific values to them.

> Let the given number *a* be separated into *x* and *y* so that the product of *x* and *y* is given as *b*. Moreover, let the square of *a* be *e*, and the quadruple of *b* be *f*. Subtract this from *e* to get *g*, which will then be the square of the difference of *x* and *y*. Take the square root of *g* and call it *h*. Then *h* is also the difference of *x* and *y*. Since *h* is known, *x* and *y* can be found.

We can write this procedure in the same abbreviated form we used for al-Khwārizmī's algorithm, but using symbols rather than numbers:

$$\text{INPUT: } a, b$$
$$\text{Square}(a) \rightarrow e$$
$$\text{Quadruple}(b) \rightarrow f$$
$$\text{Subtract}(e, f) \rightarrow g$$
$$\text{SquareRoot}(g) \rightarrow h$$

Now Jordanus observes that the problem has been reduced to the simpler one discussed previously. The two parts x and y of a are such that their difference is h, and so they can be found by the standard method for solving that problem (as given above). For example, suppose we want to find the numbers whose sum and product are 228 and 3587. By following Jordanus's instructions, and using Hindu-Arabic arithmetic, we can calculate e, f, g, and h:

a	b	e	f	g	h
228	3587	51984	14348	37636	194

So it remains to find the numbers whose sum and difference are 228 and 194, the problem that Jordanus had previously discussed. Using his method, we calculate that half of 228 minus 194 is 17, which is the smaller number, and the other is 228 minus 17, which is 211.

Nowadays we can use an electronic calculator, because it has been hard-wired to do the basic arithmetical operations. Our task is merely to supply the algorithm, which means pressing the right keys in the right order. But for many hundreds of years people would have solved such problems using the pen-reckoning methods of arithmetic, as taught by Edmund Wingate. The significance of Jordanus's exposition was that it separated the sequence of operations (the algorithm) from the individual calculations (the arithmetic). Anyone who could do the calculations could solve the problem by following the instructions carefully.

Jordanus's symbolic approach was not taken up immediately and so, for several more centuries, mathematics would continue to be written rather clumsily, in a literary style. Chapter 5 will cover some important discoveries that took place alongside the great renaissance (re-birth) of culture in the fifteenth and sixteenth centuries. But, in mathematics, the most significant advances did not happen until the seventeenth century and, when the time came, a great new domain of mathematics was born, rather than re-born.

CHAPTER 5

The End of the Middle Ages

I n this chapter you will read how people gradually became accustomed to using algorithmic methods in the later medieval period. This happened in several ways. Some of the new arithmetical procedures were adopted because they could be applied directly to complex problems that arose in the mercantile world. Other methods, both arithmetical and algebraic, suggested new problems that posed a challenge to the scholars of the time. These two threads were inter-twined, and in due course they were to form the foundation for a great new world of mathematics.

Merchants and mathematicians

The dissemination of Hindu-Arabic arithmetic in Europe began towards the end of the first millennium. It may have been partly inspired by Gerbert's reputation as a scholar, although probably not by his brief sojourn as Pope in Rome. Initially, it appears that this knowledge was confined to small groups of high-ranking clerics and officials, who had no reason to share their expertise more widely. It was to be another 200 years before the new methods began to filter through to the mercantile classes.[1]

One of the main catalysts was a book, the *Liber Abbaci*, written by Leonardo of Pisa in 1202, and revised in 1228. Leonardo is often known as 'Fibonacci', which reveals his origins as a member of a prosperous family, the House of Bonacci.[2] His father was the representative of the merchants of Pisa in North Africa, and he made sure that Leonardo was instructed in the Hindu-Arabic arithmetic. Leon-ardo also travelled with his father in the East before returning to Pisa around 1200 and writing his book. Many of the problems in the *Liber Abbaci* are taken from earlier texts, but Leonardo enlivened them with good stories. For example, here is a version of his 'Problem of the Birds', which has also been found in a manuscript written by the Islamic scholar Abu Kamil in the early part of the tenth century.

I bought 30 birds for 30 denari. Some were brown, some grey, some white, and they cost respectively 1/3, 1/2, and 2 denari. How many birds of each kind did I buy?

In the hands of a modern student this problem would be attacked by writing down some algebraic equations. Leonardo did not have these tools, but he was able to find the solution, and explain how to solve similar problems. On the reasonable assumption that only whole birds are allowed, and at least one of each kind must be bought, there is a unique answer.[3]

The *Liber Abbaci* contained instructions for carrying out all the standard arithmetical operations, using the Hindu-Arabic methods. One significant development was that Leonardo dealt with fractions in much the same way as we do today, and used the modern notation. So he was able to explain clearly how to do the awkward sums that still bother us, such as adding 17/44 and 15/26. A more subtle consequence of this approach was that people became accustomed to the idea that fractions are numbers, just like whole numbers, because they can be added and multiplied in a similar way. These operations have properties that we find reassuring: for example, just as 5 plus 3 is the same as 3 plus 5, so 17/44 plus 15/26 is the same as 15/26 plus 17/44. This viewpoint is the foundation for the modern concept of a *number-system* satisfying certain rules, and it is fundamentally different from the way of thinking that had been current in the ancient world.

The contents of the *Liber Abbaci* diffused slowly into the mercantile community, and we know that Leonardo wrote other books on the subject, some of which have not survived. His advocacy of the Hindu-Arabic system of arithmetic gradually took hold, and by the end of the thirteenth century the methods were well known not only in Pisa, but also in the flourishing Italian city-states like Genoa and Florence. Indeed, it is recorded that in 1299 the Florentines banned the Hindu-Arabic numerals, a fact that has often been misunderstood. The ban applied specifically to the use of the numerals in records, presumably so that customers who were used to the traditional Roman numerals could check their accounts. The very existence of the ban suggests that, for internal purposes, the Florentine bankers had become accustomed to using the Hindu-Arabic numerals and the associated algorithms.

Several of Leonardo's arithmetical problems had mercantile aspects, and many others, like the Problem of the Birds, could be interpreted in that way. His book contains several chapters entirely devoted to commercial arithmetic, one of which deals with the alloying of metals for the silver coinage. As far back as the third century the Romans had debased their 'silver' coins by adding significant amounts of copper, so that the coins became a form of token money. Not surprisingly, token money was unpopular. After the break-up of the Roman Empire, new self-governing nations emerged, and in due course most of them established a coinage of good silver. In England the standard coin was a silver penny, and the fineness of its alloy was maintained for many centuries, although the weight varied. But in several other places the silver coinage gradually degenerated once again, resulting in the currency of coins known as 'black money'. One such coin was the denaro piccolo of Florence, which was supposed, somewhat ludicrously, to contain one part of silver to 99 parts of copper. The value of the denaro piccolo was so small that it eventually became uneconomic to mint it (even without the trace of silver), and it was replaced by a coin known as a quattrino, also black money, but nominally worth four denari.

In due course these changes resulted in financial systems of baffling complexity. In Florence, by the fourteenth century, the coins in circulation—the money used as a medium of exchange—consisted mainly of gold florins (Figure 5.1) and base metal quattrinos. On the other hand, the money used for accounting purposes was expressed in traditional units based on the denaro and its multiples, the soldo (12 denari) and the lira (20 soldi).[4]

Back in Leonardo's time many 'silver' coins were being minted throughout Italy and the neighbouring areas. Their silver content varied considerably, which explains why the *Liber Abbaci* contains detailed instructions about how to produce alloys of specified proportions. Leonardo did not say much about gold coins, probably because they were relatively scarce when his book was first written. Very few gold coins had been minted in western Europe for several centuries, although the gold dinar (see Figure 4.6) continued its long reign as the characteristic coin of the Islamic world, and often circulated in the neighbouring Christian regions. However, during Leonardo's lifetime this situation changed significantly, and it is possible that he played some part in the process. The central figure was Frederick

Figure 5.1 European gold coins minted in the late Middle Ages. Above is an augustale of Frederick II of Sicily, and below a florin of Florence (this florin dates from 1347). Wikimedia Commons: supplied by Classical Numismatic Group Inc., http://www.cngcoins.com.

II, who became Holy Roman Emperor in 1220. Frederick had been king of Sicily since 1197, and the geographical position of that island meant that it was a meeting point for Muslims and Christians, commercially as well as intellectually. Thus it was in Sicily, from about 1231 onwards, that Frederick began to produce a new type of gold coin, known as an *augustale* (Figure 5.1). The style of the augustale was based on a classic Roman model, showing the Emperor as a military leader wearing a wreath of laurels. Frederick was a very able man, and he also took an active interest in the arts and sciences, arranging for translations of Greek and Arabic manuscripts to be made. We know that he met Leonardo in 1225, and he may well have discussed the problems of producing gold coins with him on that occasion.

Although Frederick's augustale was essentially an experiment, the time was ripe for gold coinage on a larger scale, as exemplified by the florin, which was minted in Florence from 1252 onwards. It was to be a model for gold coins in Christian Europe for several hundred years. But the model was not followed exactly, and by the beginning of the fourteenth century there were many gold coins in circulation, all with different weights and finenesses. As an example of this monetary muddle, we can imagine the difficulties of a Florentine traveller in some remote part of Europe, such as London. To pay for food and lodging he would have to exchange his florins for some locally acceptable coins. The only coins current in England at that time were silver pennies, so he would need to know the prevailing exchange rate, the number of pennies given for a florin. This number varied quite considerably (roughly in the range 30 to 42), because it depended on economic factors, such as the supply and demand for gold and silver.

One such Florentine traveller was Francesco Balducci Pegolotti, the representative of the Bardi, a leading banking house, who worked in London from 1317 to 1321. In the course of his travels he compiled a notebook, known as *La Pratica della Mercatura*, which contains an enormous amount of useful information about the moneys, weights, and measures of the places he visited. For example, the book contains a table giving the fineness of over thirty different gold coins, including the dinar, the augustale, and the florin. Pegolotti expressed the fineness of gold as a certain number of carats, much as we do today, so that pure gold is 24 carats fine. His table reports that the dinar was 23½ carats, the augustale 20½ carats, and the florin 24 carats.

The personal difficulties that Pegolotti encountered on his travels were insignificant in comparison with his professional duties, which would have included making arrangements for exporting large consignments of goods, such as wool, from London to Florence. Both the exporter and the importer expected to deal in hard cash, but sending coins on a long journey was a risky business. Furthermore, the exporter would expect to be paid in silver, the hard cash current in England, and the importer would expect to pay in gold florins, the hard cash current in Florence. To overcome this difficulty, the Bill of Exchange was developed (Figure 5.2). By this means the parties could make and receive payment in their local currencies; only the documents authorizing the payments had to be sent on a long journey.

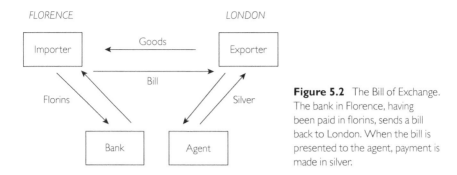

Figure 5.2 The Bill of Exchange. The bank in Florence, having been paid in florins, sends a bill back to London. When the bill is presented to the agent, payment is made in silver.

The system was kept in balance by trade in the opposite direction: in this case, goods being sent from Florence to London.

Drafting a bill of exchange required great arithmetical expertise, because the actual payments depended not only on the variable exchange rate, but also on the systems of accounting used in London and Florence. In London prices of goods like wool were expressed in marks, a mark being 160 pence. In Florence prices were expressed in lire, soldi, and denari, and at that time the gold florin was reckoned at 29 soldi or 348 denari. Such complications were the reason why Pegolotti's notebook contained a table of exchange rates, of which the following is a short extract.

A denari $34\frac{1}{4}$ sterl. il fior. viene il mar. lire 6, sol. 15, den. $5\frac{95}{137}$.

A denari $34\frac{1}{2}$ sterl. il fior. viene il mar. lire 6, sol. 14, den. $5\frac{21}{23}$.

A denari $34\frac{3}{4}$ sterl. il fior. viene il mar. lire 6, sol. 13, den. $6\frac{49}{139}$.

A denari 35 sterl. il fior. viene il mar. lire 6, sol. 12, den. $6\frac{6}{7}$.

The first line can be translated as follows.

At $34\frac{1}{4}$ sterling pence to a florin, the mark is 6 lire, 15 soldi, and $5\frac{95}{137}$ denari.

This stated value of the mark is the result of dividing the product of the numbers 160 and 348 (whose significance was explained above) by the exchange rate, giving the result, in denari, as

$$1625\frac{95}{137}.$$

In the table this amount has been converted to lire, soldi, and denari.

The complexity of the calculation, particularly division by numbers with fractional parts, suggests that the table was compiled by highly skilled arithmeticians, using the Hindu-Arabic methods described in the *Liber Abbaci*. On the other hand, a typical user of the table had a simpler task. He would have been a clerk in Florence, who needed to draw up a bill of exchange for a shipment of wool valued at 875 marks (say). This would involve ascertaining the current exchange rate, looking up the corresponding value of the mark, and multiplying by 875 to get the amount to be entered in the accounts. Such calculations are not trivial, but they can be done by any properly trained clerk. It is recorded that in 1345 over a thousand children were being taught the elements of *abaco* and *algorismo* in Florence, which suggests that arithmetical techniques, at various levels of sophistication, were much needed in commerce. The establishment of these 'abacus schools', and the existence of a group of people trained in the algorithmic arts, was to have significant consequences for the future of mathematics as a whole.

Interest upon interest, and more . . .

The main purpose of Pegolotti's notebook was to collect specific information about the coins, weights, and measures of many localities. But it also contains some general information, including tables for calculating the interest on a loan. By this time the ethical and religious objections to charging interest had been set aside; it was argued that in some respects money is a commodity like a plot of land, and hence there is ample justification for making a charge for lending it. Not only was it acceptable to charge interest on a loan of money, it was also acceptable to charge 'interest upon interest', or what we now call *compound interest*. For example, when 100 lire is lent at the annual rate of 5 per cent per year, the interest due at the end of the first year is 5 lire, so the loan effectively becomes 105 lire. At the end of the second year the interest due is 5 per cent of that amount, which comes to 5 lire 5 soldi. So the loan is 110 lire 5 soldi, which is slightly more than would have been the case if simple interest had been charged. Pegolotti's tables gave the compound interest on a loan of 100 lire for periods from 1 to 20 years, at annual rates up to 8 per cent. The calculations were tiresome, but not difficult for a competent arithmetician. For example, charging interest at the annual rate of 5 per cent is the same as adding one-twentieth of the amount at the end of each year. So for the first and second years the calculations are (in modern notation):

$$100 \times \left(1 + \frac{1}{20}\right) = 105, \quad 100 \times \left(1 + \frac{1}{20}\right) \times \left(1 + \frac{1}{20}\right) = 110\frac{1}{4}.$$

The expression

$$1 + \frac{1}{20}$$

is known as a *binomial*, signifying that it is the sum of two numbers. The figures for the subsequent years can be calculated by repeatedly multiplying by this binomial term: for example, after five years the amount outstanding (in lire) has grown to

$$100 \times \left(1 + \frac{1}{20}\right) \times \left(1 + \frac{1}{20}\right) \times \left(1 + \frac{1}{20}\right) \times \left(1 + \frac{1}{20}\right) \times \left(1 + \frac{1}{20}\right).$$

After some tedious calculations the answer is found to be 127 lire, 12 soldi, 7 denari. Faced with a lot of calculations like this, even a competent arithmetician might well ask for a simpler method. Surely the product of the five binomial terms can be simplified in some way?

Repeated multiplication of binomials occurs in several mathematical contexts, and we have already met the simplest case, the rule for squaring a binomial, in geometrical disguise. In Figure 5.3 the sides of a square have been divided into two parts, with lengths a and b so that the square itself is divided into four parts, with areas $a \times a$, $a \times b$, $b \times a$, and $b \times b$.

In modern notation we denote these terms by a^2, ab, ba, and b^2, and we take it for granted that ab is the same as ba. It follows that the area of the whole square is

$$\left(a + b\right)^2 = a^2 + 2ab + b^2.$$

This formula can be used in many ways. For example, we can apply it to the calculation of compound interest at 5 per cent over two years. The result is

$$\left(1 + \frac{1}{20}\right)^2 = 1 + 2 \times \left(\frac{1}{20}\right) + \left(\frac{1}{20}\right)^2,$$

where the second term represents two annual instalments of simple interest at 5 per cent, and the last term represents the 'interest upon interest'. A much older application occurs in the geometrical procedure used by al-Khwārizmī to solve quadratic equations, as in his example 'one square and 10 roots equal to 39' (see Figures 4.9 and 4.10). There the formula was applied to a square whose sides are divided into two parts, 5 and the unknown number. According to the algorithm the area of the whole square is $25 + 39 = 64$, so the sides have length 8 (the square root of 64), and it follows that the unknown number is $8 - 5 = 3$.

That example was easy, because the square root of 64 is a whole number. But what if al-Khwārizmī had set the example 'one square and ten roots equal to 40', so that the area of the whole square is $25 + 40 = 65$? In that case the algorithm requires us to compute the square root of 65, and this is certainly not a whole number. An obvious strategy is to begin with 8 as an approximation and try to find a better one. For this there is a rule that goes back to antiquity.[5] However, we must take care, because this rule produces approximations that are fractions. In the light

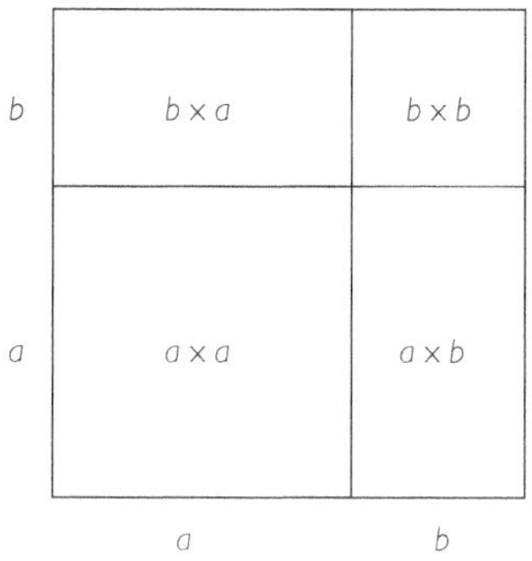

Figure 5.3 Squaring a binomial, geometrically.

of the Greek discoveries about $\sqrt{2}$ it should not come as a surprise to learn that $\sqrt{65}$ cannot be expressed exactly as a fraction, and so this method cannot give us the exact answer. Nevertheless we can use it to get as close as we please to $\sqrt{65}$, even if we cannot yet explain precisely what that number really is.

Let us pause here to review the state of the art towards the end of the fourteenth century. The Mesopotamians knew the arithmetical procedures for completing the square, and al-Khwārizmī had written them down in the form of an algorithm. It was clear that the main step was finding the square root of a certain number. The work of Leonardo, on the algebraic manipulations and the arithmetical procedures for calculating square roots, was becoming widely known. Taking a broader view, and remembering the work of Jordanus (as described in Chapter 4), we can see the outlines of the modern symbolic approach beginning to take shape. The scholars of the late Middle Ages did not use our notation, but their lines of thought were similar to the ones we use today. The leading mathematicians, blessed with great insight, would have had no difficulty in following some of the better textbooks on modern algebra.

The preceding remarks mean that we can now begin to use modern notation for the purposes of explanation: indeed, we have already started to do so. But we must bear in mind the unhistorical aspects of this procedure. When we speak of an 'equation', we must remember that the sign ($=$) for equality was not introduced until the middle of the sixteenth century. Before that, equations were still being stated in words, although abbreviations like p for plus and m for minus were common. The unknown quantity whose value satisfied the equation was referred to as the 'thing', or 'cosa' in Italian. The use of letters like x and y for unknowns, and a and b for given numbers, did not really take hold until the very end of the sixteenth century. Our modern signs like $+$, $-$, and $\sqrt{}$ were introduced sporadically, and they became standard only when the printers of books were persuaded to add them to their trays of type. In Chapter 6, we shall describe how the mastery of symbolic algebra opened up a whole new world of mathematics, but first we must explain how some new algebra was discovered using old notation.

Solving equations

Several fourteenth-century manuscripts contain problems on compound interest, apparently intended for use in the Italian abacus schools. One such problem asks for the rate of interest that will make 100 lire into 150 lire after three years. Pegolotti had tabulated the final amount when the rate is given; here we have the reverse situation, where we are asked to find the rate, given the final amount. In other words, we must find the multiplier that when applied three times to 100 gives the answer 150. Nowadays we should denote the unknown interest rate by x per cent, and write the problem symbolically in the form

$$100 \times \left(1 + \frac{x}{100}\right)^3 = 150.$$

This is a typical problem of algebra: the condition is an equation, and the answer (the rate of interest x) is an unknown. The format is very like that used by al-Khwārizmī 500 years earlier, but his equations were quadratic (involving the second power x^2), whereas here we have a cubic (involving x^3). Our modern notation makes it plain that this particular example has a very simple form, because there is essentially only one power involved.

When this problem was posed, around 1395, arithmetical methods of finding cube roots were known to the experts. The methods were very complicated, but basically they relied on a rule for making successive approximations, like the one for finding square roots. But the experts had also discovered that cubic equations present a new kind of difficulty. Whereas any quadratic equation can be solved by finding one square root, it is not generally true that a cubic equation can be solved just by finding one cube root, and that is why the compound interest problem discussed is not typical.

The legacy of the *Liber Abbaci* was a steady growth of activity in all things mathematical. In 1494, nearly 300 years later, the state of the art was represented by one of the earliest printed books on mathematics, the *Summa de Arithmetica, Geometrica, Proportioni et Proportionalita* of Luca Pacioli. This book also covered algebra and commercial arithmetic, and significantly it asserted that a general method of solving cubic equations, analogous to that for quadratics, had not yet been found.[6]

Pacioli's assertion was not expressed in the terms we use now, because our mathematical language was not available to him. However, the essence of the problem would have been clearly understood by the growing band of mathematically sophisticated people who read his book. It is thought that one of them, Scipione del Ferro, professor of mathematics at the University of Bologna, was the first to make significant progress. This probably happened around 1520, but we cannot be sure exactly, because there was a culture of secrecy about mathematical discoveries in Italy at this time. It is believed that del Ferro discovered how to solve any equation that we would now write in the form $x^3 + cx = d$, with c and d being positive numbers, and it is likely that he also solved equations of the form $x^3 + d = cx$, and $x^3 = cx + d$. At the time these had to be treated separately, because only equations with positive numbers c and d were considered to be meaningful. But del Ferro did not publish his work. About 20 years later Niccolo of Brescia, known as Tartaglia, rediscovered his method, and went on to solve all three types. Tartaglia was determined to protect his discoveries, apparently because they were his only means of earning money. But it is hard to understand why he thought that the underlying method could (or should) remain secret indefinitely, especially as it had already been discovered by del Ferro. Whatever the reason, his possessive nature led to one of the most famous disputes in the history of mathematics, between Tartaglia and Hieronimo Cardano, a leading Italian scholar of the time (Figure 5.4).

Cardano was renowned for his writings on both mathematics and medicine. He knew, as did many others, that a general method of solving cubic equations had been found, and he tried to persuade Tartaglia to reveal the secret. Eventually, in 1539, after making Cardano promise not to publish it, Tartaglia gave it to him in

Figure 5.4 Tartaglia and Cardano, from the title-pages of their books.

the form of a verse. It was, in the manner of the time, an algorithm, like those discussed in Chapter 4. Loosely translated, it went as follows.

> When a cube and some things are equal to a number:
>   find two numbers which differ in that number,
>   and whose product is equal to the cube of one-third of the number of things;
>   the difference of their cube roots is the answer.[7]

Here is a simple example. Suppose the problem is to find a *thing* such that its cube and 9 *things* make 26. In this case the given numbers are 9 and 26, and the cube of one-third of 9 is 27. The first step is to find two numbers that differ by 26 and whose product is 27. Here you might be lucky, and notice that 27 and 1 will do. Their cube roots are 3 and 1 so the answer is the difference, $3 - 1 = 2$.

The algorithm is not quite as straightforward as this example might suggest. It has two parts, the first of which is to find two numbers whose difference and product are given. This problem goes back to the ancients. In the thirteenth century Jordanus had stated the algorithm explicitly for the case when the sum and product are given, and this case (where the difference and product are given) is only slightly different. The second part involves the calculation of cube roots. Tartaglia's algorithm depends on the algebraic formula for cubing a binomial,

$$\left(a+b\right)^3 = a^3 + 3a^2b + 3ab^2 + b^3,$$

in much the same way as al-Khwārizmī's solution of a quadratic equation depends on the formula for $(a + b)^2$. The sixteenth-century Italians did not have our neat notation, but they were certainly familiar with the underlying ideas. Cardano gave a geometrical proof, which can be thought of as 'completing the cube'.

In practice the calculation does not usually run as smoothly as in the example given above. Suppose you are trying to solve the equation 'a cube and 12 things equal to 32'. You must begin by finding two numbers that differ by 32 and whose product is 64 (the cube of 4, which is one-third of 12). Using the method of Jordanus, these numbers turn out to be $8\sqrt{5}+16$ and $8\sqrt{5}-16$. To get the final answer you must find the cube roots of these numbers and calculate their difference. You might be tempted to do this with an electronic calculator, and you would probably be surprised to find that the answer is 2. It is easy to check that the cube of 2 plus twelve 2's is 32, so 2 really is the solution of the equation.[8]

Cardano was able to deal with several difficulties that arise in the solution of cubic equations, including the one described in the previous paragraph. His student Ludovico Ferrari also discovered how to solve quartic equations, where the fourth power of the unknown thing occurs. These discoveries and many others were published in 1545 in Cardano's book, the *Ars Magna*, much to the annoyance of Tartaglia. He accused Cardano of breaking his promise, and was not consoled by the fact that his own work was fully acknowledged. The ensuing dispute was bitter, and continued even after Tartaglia's death in 1557. If only mathematicians had learned from this episode that secrecy cuts both ways, a great deal of futile controversy might have been avoided.

Counting and arranging

Here is another problem from the *Liber Abbaci*.

> Seven old women are going to Rome. Each of them has seven mules and each mule carries seven sacks. Each sack contains seven loaves, each loaf has seven knives and each knife has seven sheaths. What is the total number of things?

Problems of this kind belong to an ancient branch of mathematics which we now call combinatorics.[9] The Hindus, in particular, seem to have assimilated counting and arranging into their philosophical approach to life in general. Very few of their early documents have survived, but fortunately there are several reliable copies of a manuscript written by the scholar Bhāskara in the twelfth century AD. Many things in his book are not new, but it is an excellent account of the state of Indian mathematics at that time, and is written in a charming style. It is known as the *Lilāvati*, presumed to be the name of Bhāskara's daughter, who is spoken of as an 'intelligent girl', and a 'fawn-eyed child'.

The *Lilāvati* contains several problems in combinatorics. One of them concerns the four attributes of the god Vishnu, who is conventionally depicted with four arms, each holding a symbol of one of his four attributes (Figure 5.5). Bhāskara

Figure 5.5 The four attributes of Vishnu: discus (D), conch (C), mace (M), and lotus (L).

asks: 'How many are the variations of form of Vishnu by the exchange of the mace, the discus, the lotus, and the conch?'. Each variation is an arrangement of the four objects. If we denote the arrangement shown in Figure 5.5 by DCML, then the arrangement in which the discus and the conch are interchanged can be denoted by CDML, and so on. Using this notation it is fairly easy to make a list of the possibilities, as follows.

DCML	DCLM	DMLC	DMCL	DLMC	DLCM
CDML	CDLM	CMDL	CMLD	CLMD	CLDM
MDLC	MDCL	MCDL	MCLD	MLCD	MLDC
LMCD	LMDC	LCMD	LCDM	LDMC	LDCM

These are what we now call the 24 *permutations* of the symbols D, C, M, and L. The list is laid out in such a way that it is clear how the number 24 is obtained. There are 4 choices for the object in the first position, and each row corresponds to one of these choices. Within a row we can put any one of the 3 remaining objects in the second position, and then either of the 2 remaining objects in the third position. The last object must then be in the fourth position. Hence the number of permutations is $4 \times 3 \times 2 \times 1$, which is 24.

Bhāskara gave a rule for calculating the number of permutations of any number of objects: multiply the numbers starting with 1 and ending with the number of objects. So in case of the god Sambhu, who has ten arms and ten attributes, he found that the number of permutations is

$$1 \times 2 \times 3 \times 4 \times 5 \times 6 \times 7 \times 8 \times 9 \times 10 = 3628800.$$

Another problem in the *Lilāvati* deals with the number of *combinations* of a given size that can be made from a given set of objects. For example, from the four attributes of Vishnu we can make the following combinations of two:

D and C   D and M   D and L   C and M   C and L   M and L.

Note that 'D and C' is the same combination as 'C and D': the order does not matter, so there are six combinations in this case. Here is one of Bhāskara's examples, taken from a much older source.

> Mathematician, tell me how many are the combinations . . . with . . . six different flavours, sweet, pungent, astringent, sour, salt and bitter, choosing them by ones, twos, threes, and so on?

For the 'combinations' with only one of the flavours, the answer is clearly 6. At first sight, the number of combinations with two flavours seems to be $6 \times 5 = 30$, as we can choose the first one in six ways and the second (different) one in five ways. But (for example) 'sweet and pungent' is the same as 'pungent and sweet', so the required number is actually only half that: 15.

Bhāskara gave a general rule for the number of combinations in ones, twos, threes, and so on. In the case of six objects he described his rule in terms of the array

6	5	4	3	2	1
1	2	3	4	5	6.

For example, the number of combinations when the objects are taken in threes is obtained by finding the product of the first three numbers in the top line and dividing by the product of the first three numbers in the bottom line:

$$\frac{6 \times 5 \times 4}{1 \times 2 \times 3} = 20.$$

Using this rule, you can check that the numbers of combinations are

size:	1	2	3	4	5	6
combinations:	6	15	20	15	6	1.

If you are wondering why this topic has been introduced here, the reason is that it has a close connection with the algebra of binomials, which played such an important part in the solution of equations. To make this connection clear, we need a little modern notation. Nowadays the number of combinations of n objects, taken r at a time, is written as

$$\binom{n}{r},$$

and we say it as 'n choose r'. Bhāskara's rule for six flavours gave us the answers

$$\binom{6}{1} = 6, \quad \binom{6}{2} = 15, \quad \binom{6}{3} = 20,$$

and so on. Bhāskara gave several other examples in the *Lilāvati*, and clearly he knew the general rule for the number of combinations of n objects taken r at a time, which we now express by the formula

$$\binom{n}{r} = \frac{n \times (n-1) \times \ldots \times (n-r+1)}{1 \times 2 \times \ldots \times r}.$$

The link between these 'combination numbers' and the algebra of binomials is simple enough, given the right notation. But in the absence of algebraic notation the historical threads are entangled. We have explained how the formulae

$$(a+b)^2 = a^2 + 2ab + b^2, \quad (a+b)^3 = a^3 + 3a^2b + 3ab^2 + b^3,$$

formed the basis of the algorithms for solving quadratic and cubic equations, and for finding square roots and cube roots. For that reason the numbers 1 2 1 and 1 3 3 1 eventually became known as *binomial coefficients*. When the medieval arithmeticians went on to discover methods for calculating fourth roots and fifth roots, they found that the relevant numbers are

$$1 \quad 4 \quad 6 \quad 4 \quad 1 \quad\quad \text{and} \quad\quad 1 \quad 5 \quad 10 \quad 10 \quad 5 \quad 1.$$

So a remarkable pattern began to emerge.

```
              1    1
           1     2     1
        1     3     3     1
     1     4     6     4     1
  1     5     10    10    5    1
```

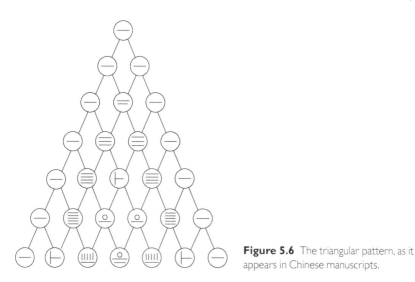

Figure 5.6 The triangular pattern, as it appears in Chinese manuscripts.

In this array each number is the sum of the numbers to the left and right of it in the line above. Surely this cannot be an accident? If you apply the same rule to calculate the next row, you will get

$$1 \quad 6 \quad 15 \quad 20 \quad 15 \quad 6 \quad 1.$$

These are the numbers that arise from Bhāskara's rule for the combinations of six objects, as given above; they are also the coefficients that occur in the sixth power of a binomial,

$$(a+b)^6 = a + 6a^5b + 15a^4b^2 + 20a^3b^3 + 15a^2b^4 + 6ab^5 + b^6.$$

Now you can see what is going on: *the binomial coefficients are the combination numbers.* The rule for forming a new row of binomial coefficients can be justified by using a simple property of the combination numbers.[10]

The remarkable triangular pattern formed by the combination numbers has been found in Chinese manuscripts from the thirteenth century (Figure 5.6). It is also found in Islamic manuscripts. We shall probably never know its ultimate origins, and it was not until algebraic symbolism became available that its relationship with the binomial coefficients was fully understood. Nowadays it is often known as Pascal's Triangle, after the seventeenth-century French mathematician who was the first to publish an account of it in modern terms. But he was neither the first nor the last to obtain startling mathematical results from this memorable array of numbers.

How to keep a secret: the medieval way

The art and science of combinatorics has many applications. It developed in a rather haphazard way, often in response to questions that were not, apparently, of great mathematical significance. One such problem was the age-old problem of sending secret messages. For example, the Roman author Suetonius tells us that Julius Caesar was in the habit of sending written instructions to his army commanders. Unfortunately, the assumption that his enemies were illiterate proved to be false, and when the barbarians intercepted a message they were able to make sense of it. So Caesar had to devise a way of concealing his instructions, while ensuring that his commanders could understand them. This is what we loosely refer to as a secret code or, in modern terminology, a cryptosystem.[11]

Alphabets like the Greek and Latin ones were used in many ingenious methods of trying to ensure secrecy. Some of these methods were not very effective: for example, the method of writing each word backwards, so that ATTACK becomes KCATTA. Caesar's cryptosystem was rather more effective. According to Suetonius he replaced each letter by the one that comes three places later in the standard alphabetical order, so that ATTACK becomes DWWDFN. One advantage of this method is that any number, not just three, can be used. The chosen number is known as the *key*, and Suetonius says that Caesar's nephew Augustus used one as the key, instead of three. It is quite possible that Caesar himself changed the key regularly; if he used a different key for each day of the week, then his instructions would have been more secure.

Sadly, Caesar's method has a fatal weakness. We should not blame him too readily, because over the past two millennia all cryptosystems have been found to have their limitations. For this reason there has been a constant battle to make better ones, driven by the discovery that any new system is destined to fail eventually. As we shall see, mathematical ideas have come to play an increasingly important part in this battle.

Let us begin by looking more closely at Caesar's cryptosystem. Using our English alphabet with 26 letters his method was to choose a number k between 1 and 25 and replace each letter by the one that is k places later, in alphabetical order. This rule extends in an obvious way to the letters at the end of the alphabet; for example, when k is 5 the letters

A B C D E F G H I J K L M N O P Q R S T U V W X Y Z

are replaced by

F G H I J K L M N O P Q R S T U V W X Y Z A B C D E.

Using this rule the sender would encrypt the message SEE YOU TOMORROW as XJJ DTZ YTRTWWTB. The intended receiver is presumed to know the key, and so it is possible to decrypt the message by using the rule in reverse. On the other hand, if someone who does not know the key intercepts the message, decoding is not so easy. The main lesson from history is that it is futile to try to conceal the method of encryption: the security of the system must depend only on the choice of the key. So let us imagine that you, Barbara/Barbarus, have intercepted an encrypted message such as SGZNY OY MUUJ LUX EUA. You know what system is being used, so your problem is to find which key was used for this particular message. As the only possibilities are the numbers from 1 to 25, you would try these keys in turn, remembering that if the key is k, you must go back k places in the alphabet to find the original message. As a start, try

$$k = 1, \text{giving RFYMX NX}\ldots,$$
$$k = 2, \text{giving QEXLW MW}\ldots.$$

Clearly the key is not 1 or 2 because, if it were, the original message would not make sense; there is no need to work through the whole message to establish this fact. So you must continue to try the keys $k = 3, 4, \ldots, 25$, until a meaningful message is found.[12]

The method of trying all the keys in turn is known as Exhaustive Search. It is successful in the case of Caesar's system, because the number of keys is small, but it might fail for practical reasons if the number of keys were much larger. So it is significant that the system can easily be modified to achieve that aim. The rule for substituting letters does not have to be quite as simple as Caesar's; for example the word PERSONALITY defines a permutation in which the letters

A B C D E F G H I J K L M N O P Q R S T U V W X Y Z

are replaced by

P E R S O N A L I T Y B C D F G H J K M Q U V W X Z.

This kind of rule is useful because the sender and receiver can memorize the keyword: it does not have to be written down. But any permutation of the letters will do, provided there is a secure means of sharing it between the sender and the receiver. So the number of possible keys is large: in fact it is the number of permutations of the 26 letters which, according to Bhāskara's rule, is $26 \times 25 \times 24 \times \ldots 3 \times 2 \times 1$. This turns out be a truly huge number, far greater than the number of microseconds since the Earth was formed, and so the attack by exhaustive search is beyond the capabilities of us barbarians.[13]

'He that hath a secret to keep must keep it secret that he hath a secret.' This is an old saying that applies with special relevance to the activities of those who make cryptosystems and those who attack them. For this reason the historical record is very patchy. However, we can infer that cryptosystems based on arbitrary permutations of letters were in frequent use by about 850 AD, because there is documentary evidence from that time. This evidence comes from the Islamic world, where a number of relevant Arabic manuscripts have been discovered.[14] The principal one came to light only in 1987. It was written by the philosopher al-Kindi (801–873), who spent most of his life in Baghdad writing literally hundreds of books. His book on cryptosystems contains many ideas that are still relevant today, and in particular he describes a general method of attacking the system based on substitution of letters.

Al-Kindi's method is based on the observation that, in any written message, the letters occur with different frequencies. For example, in modern English the letter E occurs very often, whereas J is relatively rare. Furthermore, it turns out that the frequencies of the letters are fairly constant, whatever the source of the message. Typical values of these frequencies, expressed as the number of occurrences per 10000 symbols in English, are tabulated below.

A	B	C	D	E	F	G	H	I	J	K	L	M
765	128	289	418	1243	296	143	582	669	8	41	370	269
N	O	P	Q	R	S	T	U	V	W	X	Y	Z
748	794	237	7	674	650	901	300	109	155	29	166	9

This table is the basis of the attack by the method known as frequency analysis. In any message of reasonable length the symbols will occur with frequencies close to those given in the table. This means that the encrypted forms of some letters can be identified fairly easily. For example, it is likely that E is the most frequently occurring letter in the original message, and so the letter that occurs most frequently in the encrypted message is likely to be the one that represents E. The representatives of other frequently occurring letters can be inferred in the same way, although some trial-and-error may be needed. In summary, the attack by frequency analysis is carried out in three stages.

1 Count the frequencies of the symbols in the encrypted message.
2 Compare these data with the frequencies given in the standard table.
3 Test the likely correspondences of the symbols until a meaningful message is found.

This method of using data to justify a practical conclusion is typical of what we now call *statistics*. Of course, the reliability of the conclusion depends on two factors, the accuracy of the data and the structure that underlies the data. The collection of data for administrative purposes is a very old practice, but its application to encrypted messages was certainly one of the earliest instances of a glimmer of a mathematical structure.

The writings of al-Kindi and others confirm that the theory and practice of cryptosystems was known in the Islamic world before the end of the first millennium. Gradually that knowledge must have filtered through to the Christian countries, probably as a result of war, rather than trade. During this period the makers of cryptosystems were mainly concerned with defending against the attack by frequency analysis. Many ways were suggested of doing this, but most were impractical because they were too complicated: if the intended receiver cannot decrypt the message fairly easily, the system is unworkable. However, in the fifteenth century a more successful approach emerged. The basic idea is to set up a system in which any given symbol, such as E, is not always encrypted in the same way. This can be achieved by using a rule in which the key changes according to a rule known only to the sender and receiver.

One of the first systems of this kind was invented in 1467 by Leon Battista Alberti, an Italian known for his writings on architecture and the theory of

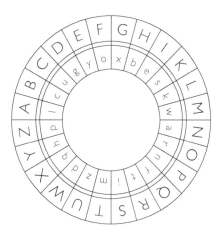

Figure 5.7 Alberti's disc. The rings rotate independently, and each position determines a key for encryption.

perspective. He proposed that the sender and the receiver should have identical copies of a disc with two concentric rings, like the one shown in Figure 5.7. In this example each ring contains 24 letters; the missing letters are presumed to occur rarely, if at all. The two rings can be rotated relative to one another, and setting them in a given position corresponds to choosing a key. With this device it is easy to design procedures in which the key changes in the required manner: for example, the inner ring could be rotated clockwise by one segment after every five symbols.

Another method, similar to Alberti's but requiring no special equipment, was suggested by Giovanni Belaso in 1553. It is usually named after Blaise de Vigenère, who discovered essentially the same method some years later. It is based on Caesar's method of shifting each letter by a certain number of places, and allowing this number to vary according to the position of the letter in the text. The various shifts are determined by a keyword. If the keyword is CHANGE, the shifts are 3, 8, 1, 14, 7, 5, corresponding to the positions in the alphabet of the letters C, H, A, N, G, E. Using this keyword, we split the message into blocks of six letters, and apply the shift of 3 to the first letter in each block, the shift of 8 to the second letter, and so on.

The Vigenère system has the effect of smoothing out the frequencies of the letters in the encrypted message. For example, suppose that the original message contains letters with their usual frequencies, as given in our table, and the keyword is CHANGE. Then the letter A in the encrypted message represents one of the letters X S Z M T V, because these are the letters that come 3, 8, 1, 14, 7, 5 places before A in the alphabet. So we can expect A to occur about

$$\frac{1}{6}(29+650+9+269+901+109)$$

times in every 10000, which is about 358. When we work out the expected frequencies for the other letters in the same way, we find that all the answers are fairly close. In other words, the irregularities that make frequency analysis possible have been flattened out. For several centuries the Vigenère system was thought to be unbreakable, and was referred to as *le chiffre indéciffrable* (the undecipherable cipher). But, as we shall see at the end of our story, like all cryptosystems it was eventually broken. The breakthrough was the result of using more sophisticated statistical methods, supported by advances in the language of mathematics.

CHAPTER 6

A New World of Mathematics

After many centuries of obscurity, the seventeenth century was a time of great clarification in the theory and practice of mathematics. As an educated person living at that time, you would have been able to find out about many of the new ideas by reading books written in everyday language, rather than Latin. Consequently you would be aware that advances in mathematics were finding applications in many fields of human activity, as well as helping us to understand the great wonders of the universe.

Measurement and calculation

In 1584 Queen Elizabeth sent Sir Walter Raleigh to the New World, with the intention of extending England's colonial territories.[1] Raleigh's first problem was how to get there (and back) safely. Eventually he sought help from a young Oxford graduate, Thomas Harriot, who turned out to be a mathematical genius. Harriot studied the mathematical theory of map projections, designed instruments for making observations, and accompanied Raleigh on one of his voyages.

Navigation at sea depended on taking measurements of several kinds, and using them as inputs to a calculation that was intended to determine the ship's position. The branch of geometry that we call *trigonometry* was the foundation for these calculations.[2] But in the sixteenth century navigation was not an exact science: an officer on one of Raleigh's early voyages reported that the navigators had often disagreed by up to a hundred miles. There were maps with indications of latitude and longitude, but unfortunately the maps were of little use in the middle of the Atlantic Ocean. The magnetic compass was used for deciding what bearing to follow, but here too there were many difficulties. Thomas Harriot studied these matters in mathematical terms, and he was able to turn his results to practical advantage. One of his contributions was to make tables of what he called 'meridional parts', which told the navigators how to make a more precise estimate of the required bearing. The production of these tables required a great deal of calculation, and in this respect he was also able to make significant improvements. Later, his methods were to be applied to arithmetical tables more generally.

By the end of the sixteenth century, arithmetic was also playing an important part in more mundane affairs. The financial transactions involved in commerce had become increasingly complicated: the period of a loan was often measured in weeks or days, partial payments or repayments were allowed, and the rate of interest varied. The use of financial instruments such as leases and annuities created similar difficulties. Consequently there was a demand for skilled arithmeticians—in particular, for people who could make reliable tables to assist those who had to do the calculations. One such work was the *Tafelen van Interest* (1582) written by Simon Stevin, an accountant and engineer who lived in Bruges. The same author's *De Thiende* (1585) was a work of a different kind. In this book he advocated the use of decimal fractions, and explained why they were superior for computational purposes to the 'vulgar' fractions, like 17/44, then in use.[3] Although Leonardo of Pisa had given the rules for calculating with vulgar fractions back in 1202, it is safe to say that few people understood the reasons behind them. That was hardly surprising, as the textbooks of the day offered little or no explanation. The extract shown in Figure 6.1 is typical.

Stevin suggested that it would be simpler, particularly for the users of tables, if all numbers were represented with the same power of ten underneath. When that is done, the addition of fractions is no more difficult than the addition of whole

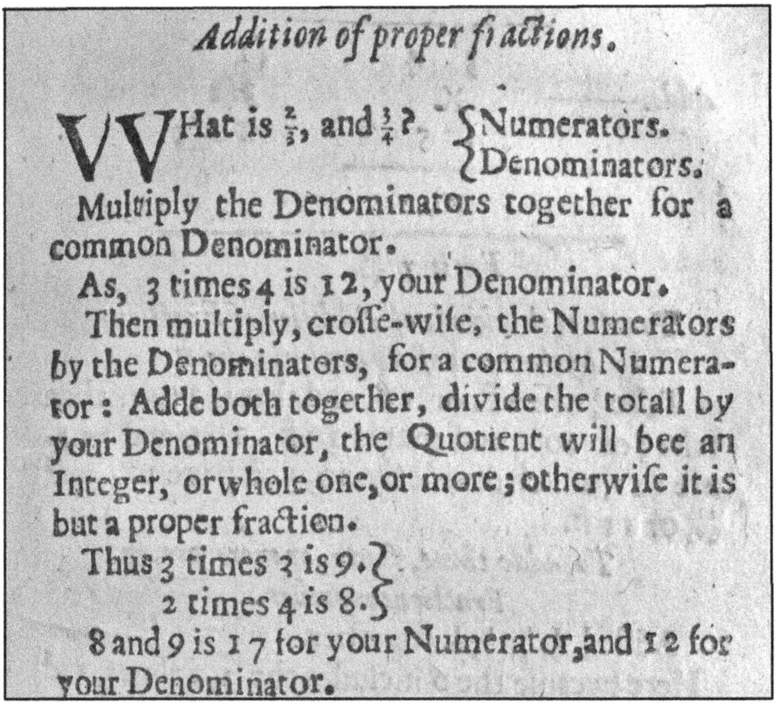

Addition of proper fractions.

WHat is ⅔, and ¾? {Numerators. {Denominators.
Multiply the Denominators together for a common Denominator.
As, 3 times 4 is 12, your Denominator.
Then multiply, crosse-wise, the Numerators by the Denominators, for a common Numerator: Adde both together, divide the totall by your Denominator, the Quotient will bee an Integer, or whole one, or more; otherwise it is but a proper fraction.
Thus 3 times 3 is 9.}
    2 times 4 is 8.}
8 and 9 is 17 for your Numerator, and 12 for your Denominator.

Figure 6.1 How to add fractions, according to Nicholas Hunt's *Hand-maid to Arithmetick* (1633).

numbers, and the rule is self-evident. For example, the numbers 17/44 and 15/26 can be represented (approximately) by

$$\frac{386363}{1000000} \quad \text{and} \quad \frac{576923}{1000000},$$

and their sum can be found simply by adding 386363 and 576923. The answer is 963286, so the sum is

$$\frac{963286}{1000000}.$$

The notation was subsequently improved by the use of the decimal point, so that the answer is now written as 0.963286. More digits can be used if a better approximation is needed. Several other mathematicians besides Stevin had made similar suggestions; indeed the idea can be found in the works of al-Uqlīdisī, as described in Chapter 4. But Stevin's book was the catalyst, and decimal fractions eventually became the standard tool of arithmetic.

The fact that the arithmetical algorithms for multiplication and division are significantly more complicated than those for addition and subtraction was also the stimulus for another invention which was to revolutionize the art of calculation. The idea is quite simple. To every number we assign another, which we call its *logarithm*. These logarithms are calculated, and are made generally available in a table. To *multiply* two given numbers, we simply look up their logarithms in the table, and *add* them. The answer is the number whose logarithm is this sum. To divide one number by another, we subtract the logarithm of the second number from the logarithm of the first. Many calculations of this kind were required in the construction of commercial tables. For example, Pegolotti's work on exchange rates, discussed in Chapter 5, contained a calculation that involved multiplying 348 by 160 and dividing the result by 34. With a table of logarithms, the problem is reduced to working out

$$\log(348) + \log(160) - \log(34).$$

The required answer is the number whose logarithm is the result of this calculation, and this can be found in the table.

The first person to devise a practical system of this kind was a Scottish nobleman, John Napier. His table of logarithms was published in 1614. It was intended specifically for some trigonometrical calculations, and the system was quite complicated, but to Napier goes the credit for the original invention, and for the word logarithm. The creation of a more practical system came a few years later. Henry Briggs, Professor at Gresham College in London, discussed the problem with Napier, and they agreed on some technical details. They decided that, for any numbers *a* and *b*, the logarithms of *a*, *b*, and *ab* should satisfy the basic rule

$$\log(ab) = \log(a) + \log(b).$$

But this condition is not quite enough to define the logarithm uniquely: a little more information is needed. As multiplying any number by 1 has no effect, the number 1 should be represented by adding 0. In other words, the logarithm of 1 should be 0. But what number should be chosen so that its logarithm is 1? As the decimal notation for numbers and fractions was now being widely used, Napier and Briggs agreed that the obvious choice was 10. That was the origin of what became known as *common* logarithms.

Compiling tables of common logarithms was hard work, but their usefulness for calculations in navigation and astronomy was immediately recognized, and by the end of the 1620s such tables were available for numbers up to 100000. One further improvement was needed. As noted above, the last step in a calculation with logarithms is to find the number whose logarithm is the result of adding (or subtracting) the logarithms of the various numbers involved. This can be done by scanning the table of logarithms in reverse, but it is much more convenient to use a table drawn up specifically for that purpose—a table of so-called *anti*logarithms. Consequently, in the 1640s a reliable table of antilogarithms was compiled, using methods originally developed by Thomas Harriot in his work on navigation. For over 300 years these tables of logarithms and antilogarithms were in constant use wherever arithmetical calculations were needed, in science, industry, and commerce.

To the limit

Stevin's work on decimals was read by Thomas Harriot. His successful association with Sir Walter Raleigh had continued into the 1590s, after which he was employed by the Earl of Northumberland. In these pleasant circumstances he was able to pursue his mathematical interests untroubled by financial worries, and he made good use of his time. On 26 July 1609 he was the first person to make a drawing of the moon as seen through a telescope, beating Galileo by four months. He also produced copious notes on a variety of mathematical topics. His comfortable situation meant that he did not need to write up his work formally, and none of it was published in his own lifetime. Fortunately, many of his private papers survived, and they show that he was surely one of the leading scholars of his day.[4]

Among Harriot's papers there is one (Figure 6.2) that deals with compound interest. The practice of charging interest upon interest had been common in Pegolotti's time, and Harriot made a new contribution which is highly significant, not just for its specific content, but because it foreshadowed several of the most important mathematical discoveries of the seventeenth century. As can be seen, Harriot himself provided almost no explanation of his working, but with hindsight we can understand what he was doing.

His motivation was the observation that if interest is added more often than once a year, but at the same equivalent rate, then the yield will be greater. He denoted the annual rate of interest by $1/b$ (for example $b = 20$ corresponds to 5 per cent),

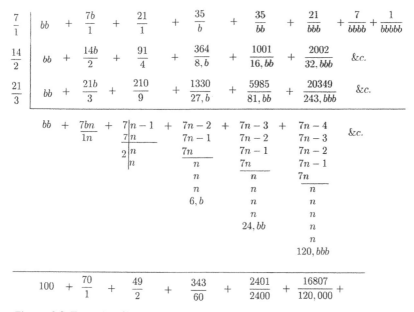

Figure 6.2 Transcript of Harriot's notes on a problem in compound interest. © The British Library Board, Add Ms 6782 f.67.

so one pound invested for one year will yield $1+1/b$ pounds. But if the interest is added twice yearly at the rate of $1/2b$ then, according to the binomial formula, the yield at the end of one year is

$$\left(1+\frac{1}{2b}\right)^2 = 1 + 2 \times \left(\frac{1}{2b}\right) + \left(\frac{1}{2b}\right)^2.$$

You should convince yourself that this is slightly more than $1 +1/b$. If the interest is added three times a year, at the rate $1/3b$, then the yield is $(1+1/3b)^3$, which is greater still, and so on. The question is: what happens when the compounding is done with greater and greater frequency, but at the same equivalent rate? In particular, can the yield become arbitrarily large?

To attack this problem Harriot considered what happens when the interest is added n times a year, at the equivalent rate of $1/nb$. Suppose the normal practice of adding interest once a year is followed, so that n is 1, and the initial loan is b^2 pounds. Then the yield after seven years is given in the first line of Harriot's calculation: we would write it as

$$b^2\left(1+\frac{1}{b}\right)^7 = b^2 + 7b + 21 + \frac{35}{b} + \frac{35}{b^2} + \frac{21}{b^3} + \frac{7}{b^4} + \frac{1}{b^5}.$$

The numbers 1, 7, 21, 35, 35, 21, 7, 1 are the binomial coefficients. It is clear that Harriot was familiar with these numbers; indeed the algebraic formula for calculating them appears in several of his papers, including this one.

The second and third lines of Harriot's manuscript give the result of applying the binomial formula to the cases $n = 2$ and $n = 3$. He then considered what happens when n is allowed to become arbitrarily large: this is what we now refer to as *continuous compounding*. For the specific value $b = 10$ he obtains the series of terms

$$100 + \frac{70}{1} + \frac{49}{2} + \frac{343}{60} + \frac{2401}{2400} + \frac{16807}{120000} + \ldots .$$

It seems that Harriot had calculated a few more terms, and noticed that they quickly become very small. So he converted his terms into pounds, shillings, and pence, and added them up, concluding that 100 pounds invested at 10 per cent for seven years, with continuous compounding, will yield approximately 201 pounds, 7 shillings, and 6.06205 pence. The significant point is that the yield does not become arbitrarily large: even though there are infinitely many terms, their sum has a definite limit, which can be calculated to any required degree of accuracy. Harriot himself made the cryptic comment 'not 7/100', suggesting that he had estimated the sum of the remaining terms, and was confident that the true value of the total did not exceed 201 pounds, 7 shillings, and 6.07 pence. From a theoretical point of view, that was a very important remark. As we shall see, infinite series like this arise in many other contexts, and mathematicians need to develop reliable methods of dealing with them.

Algebra and geometry combine

The introduction of symbolic notation led to a clearer understanding of arithmetical procedures, such as Harriot's summation of the terms in his compound interest problem. By about 1630 it was common practice to represent an unknown quantity by a letter, such as x, and to use the signs + (plus), − (minus), and = (equals), much as we do now. The rules for operating with these symbols were by then well understood, and the time was ripe for a further extension of algebraic methods—in geometry.

The central figure was René Descartes, born in France in 1596. Around 1619 he resolved to devote his life to the study of the foundations of knowledge. Much of his work was philosophical in nature, and many books have been written about it. But he also studied mathematics, and his famous *Discours de la Méthode* (1637) contained a large appendix entitled *La Géométrie*. In this work Descartes showed how algebra and geometry could be united, using algebraic symbols to represent geometrical quantities. This discovery has enormous implications, although the basic idea was not, in itself, particularly new. Just as the position of a ship at sea can be specified by its latitude and longitude, so the position of a point on a plane can be specified by two numbers, say x and y. In due course it became conventional to define x and y as the distances of the point from two lines intersecting at right angles, and nowadays the two lines are usually oriented so that the horizontal one

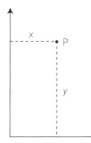

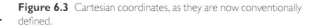

Figure 6.3 Cartesian coordinates, as they are now conventionally defined.

(the *x*-axis) points to the right, and the vertical one (the *y*-axis) points upwards (Figure 6.3). In honour of Descartes, the numbers *x* and *y* are known as Cartesian coordinates. Mathematically, there is no distinction between a point P and its coordinates (x, y).

Descartes and others went on to study what happens when a point with coordinates (x, y) is allowed to vary in such a way that *x* and *y* satisfy a relationship defined by an algebraic equation. For example, when *m* is a fixed number, an equation of the form $y = mx$ means that *y* is a constant multiple of *x*. In this case the point (x, y) will lie on a straight line, typical points being $(0,0)$, $(1, m)$, $(2, 2m)$, and so on, as shown in Figure 6.4 (1).

More generally, an equation linking *x* and *y* will define a curve, consisting of the points whose coordinates satisfy the equation. Many useful curves can be defined in this way: for example, to define a circle, we can use the fact that all points on the curve are at the same distance *r* (the radius) from the centre. Taking the centre to be the point $(0,0)$, and (x, y) as a typical point on the circle, the Pythagorean Theorem tells us that $x^2 + y^2 = r^2$, and this is therefore the equation that defines the circle, as shown in Figure 6.4 (2).

The prospect of using algebraic methods to study properties of curves in general was obviously a strong motivation for Descartes. The curves studied by the Greeks and their successors had been defined geometrically, but now curves could

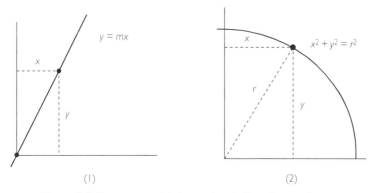

(1)   (2)

Figure 6.4 Two curves and their equations in Cartesian coordinates.

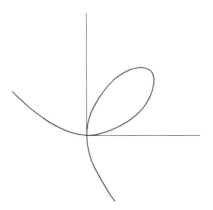

Figure 6.5 The folium of Descartes.

be defined algebraically. One of the new curves studied by Descartes was defined by the equation $x^3 + y^3 = 6xy$; it is now known as the *folium of Descartes*. At that time, it was not obvious how to set about sketching this curve. The standard method (still taught in schools) of picking some values of x and working out the corresponding values of y is almost useless, because the resulting equation for y is generally a cubic. For example, the y-coordinate of a point with x-coordinate equal to 1 must satisfy $1 + y^3 = 6y$. Descartes himself probably began by trial-and-error, finding a few points that lie on the curve, such as (0,0), (3,3), (4/3, 8/3), and (8/3, 4/3). After that, things got sticky. However, he was able to convince himself that, for positive values of x and y, the curve is a loop, as shown in Figure 6.5. But he seems to have got the big picture slightly wrong, because he thought that the whole curve comprised four similar loops, resembling a symmetrical four-leaved flower. In fact, there is only one loop, and the correct picture is the one shown.[5]

Descartes seems to have used the folium as a test case for his programme of solving geometrical problems by algebraic methods. One such problem, with many applications, was the construction of a straight line that is *tangent* to a curve at a given point: that is, a line that touches the curve but does not cross it. In the case of a circle, the construction is easy, because the tangent line is perpendicular to the radius line at the given point, and so it can be constructed by the ruler-and-compasses methods of the classical geometers. However, that is a very special case. Descartes discovered a more general algebraic method of finding tangents, but it was very complicated, and around the time (1637) that he was writing *La Géométrie* he was told that a better method had been invented by Pierre de Fermat (1601?–1665), a remarkable mathematician about whom we shall have a lot more to say in due course. Initially Descartes was not impressed by Fermat's method for finding tangents, because he thought that it could be used only in certain special circumstances. So he challenged Fermat to apply the method to the folium, fully expecting that he would fail. But Descartes was wrong: Fermat solved the problem almost immediately and Descartes had to apologize, which he did rather grudgingly.

The folium is also a good test case for another general problem, much studied in the seventeenth century. What is the area enclosed by the loop of the folium? The remarkable story of how this problem was solved is our next topic.

Back to the future: Archimedes

The problem of measuring an area bounded by curved lines presents great difficulties. The Greeks had failed to square the circle: that is to say, they could not find a ruler-and-compasses construction for a square equal in area to a given circle. That problem is in fact unsolvable. However, Archimedes had successfully pioneered a new approach to such problems, and we are now ready to consider one of his greatest achievements.

It concerns the area of a region whose boundary is formed by a straight line and part of another curve, a *parabola* (Figure 6.6). The parabola was defined geometrically by the Greeks, and many of its properties had been established, so that Archimedes was able to use them in his calculation. His method was particularly remarkable because it produced an exact answer, although no algebra was used.

Archimedes' first step was to construct a triangle with the given straight line as its base. The corners of the triangle lie on the parabola, and it has the greatest possible height subject to this constraint. The area A is simply half the base times the height. Then Archimedes constructed two more triangles in the same way, taking as their bases two sides of the first one, as shown in Figure 6.6. The crucial step is to show that the sum of the areas of these two triangles, B and C, is one-quarter of A. So the total area $A + B + C$ is equal to $5A/4$. Repeating the construction produces four more triangles, and by the same argument the sum of their areas is $(B+C)/4$, that is, $A/16$. Now the total area is

$$\left(1 + \frac{1}{4} + \frac{1}{16}\right) A = \frac{21}{16} A.$$

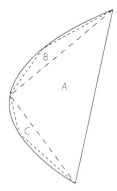

Figure 6.6 Archimedes' method for finding the area bounded by a parabola and a straight line.

Another similar step adds a term $A/64$ and the total becomes $85A/64$. It must be remembered that Archimedes did not use the Hindu-Arabic numerals, and he did not have algebraic notation to help him understand the behaviour of the sequence of approximations: A, $5A/4$, $21A/16$, $85A/64$. Nevertheless he was able to show that, as the number of terms increases, the total area approaches $4A/3$. He did not express his argument in quite that way, but his conclusion was correct: the required area is four-thirds the area of the inscribed triangle.[6]

Archimedes' work was far ahead of its time. A major difficulty was how to justify assigning the definite number $4A/3$ to the sum of an infinite number of terms.

In the later Middle Ages scholars wrote a lot about infinity, and a few of them considered the problem of finding the sum 'to infinity' of a series of terms. But the lack of a clear symbolic notation meant that most of their conclusions were obscure, and it was not until the seventeenth century that significant progress was made.

Fermat and the new algebra

In the seventeenth century there was renewed interest in the problem of calculating areas of regions bounded by curved lines. This was known as the problem of 'quadrature'. Fermat's success in solving the tangent problem for the folium of Descartes may well have encouraged him to attack the quadrature problem for that curve. We know that he eventually succeeded in finding the area of the loop of the folium, but his method was not published until 1679, long after his death in 1665. By that time many quadrature problems had been solved by other mathematicians, but Fermat had developed a powerful general method that marks an important step in the story.

Like Archimedes, Fermat used a sum of terms in his calculation and, also like Archimedes, he did not have the mathematical tools to express his findings with complete logical precision. He began by stating, in words, a theorem which he said was very well known (*notissimum*). It concerns the sum 'to infinity' of a series

$$S = a + ax + ax^2 + ax^3 + \dots,$$

where the terms form what is known as a geometric progression. This means that each term is obtained from the previous one by multiplying it by the same number, x. When x is less than 1 the terms get smaller, and if we are brave about infinity there is an easy way to derive a formula for S. Multiply through by x, getting

$$xS = ax + ax^2 + ax^3 + \dots,$$

and subtract this from the first line. On the left the result is $(1 - x)S$, and on the right all the terms except a disappear, so

$$S = \frac{a}{1-x}.$$

Archimedes' work on the parabola had depended on summing the geometric progression with $a = A$ and $x = 1/4$. He found that $S = 4A/3$, which agrees with the general formula. But Archimedes and Fermat did not have the advantage of a theory of limits, and they trod carefully in their approach to the infinite. Probably for this reason Fermat used a rather roundabout argument, but it turns out to be equivalent to the bold one given above. His contribution to the quadrature problem was to apply the formula for the sum of a geometric progression specifically to curves like the one shown in Figure 6.7.

The perimeter of the shaded region is clearly infinite, but Fermat was able to prove that in many cases its area is finite. He defined his curves by supposing that the y values GE, HI, ON, and so on, are related to the x values AG, AH, AO by an algebraic formula. Then, by making suitable choices for the x values, he showed that the areas of the rectangles GEIH, HINO, ONPM form a geometric progression with decreasing terms. The sum of this progression is known, and so the area can be calculated exactly. For example, when the equation of the curve is $y = 1/x^2$ and AG is 1, the area is 1. Fermat successfully applied this technique to many other curves of the same form, but it would not work in one specific case, the curve $y = 1/x$. This is a classic curve, known as a *hyperbola*. Like its cousin, the parabola, it had been the subject of much study since the Hellenistic period, but all attempts at quadrature had failed. As we shall see in Chapter 7, there is a rich seam of mathematics here, but its discovery had to wait until the second half of the seventeenth century.

Fermat was also responsible for another important algebraic technique. In one of his first contributions to mathematics, around 1630, he had considered the following: 'Given a line segment AB, find the position of a point P on the segment for which the product of the lengths AP and PB is greatest'. A reasonable guess is that the maximum occurs when P is the mid-point of AB, and that was certainly accepted as a fact in Fermat's time. His contribution was to suggest a general method for proving that it is indeed the right answer, using the newly fashionable

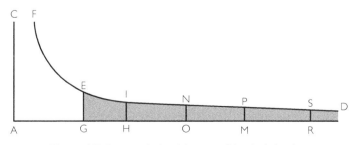

Figure 6.7 Fermat calculated the area of the shaded region.

Figure 6.8 Fermat's problem: find the point P such that $x(l - x)$ is greatest.

symbolic algebra. As in Figure 6.8, let the length AB be l and let AP be x, so that PB is $l - x$. The problem is to find the value of x for which the product $x(l - x)$ attains its maximum value. Fermat considered what happens when x is slightly altered and is replaced by $x + e$, so that the product $x(l - x)$ becomes $(x + e)(l - x - e)$. He claimed that, at the maximum point, this change is essentially zero (he used the word 'adequated'). Applying the rules of algebra, he found that the change can be written in the form $(l - 2x - e)e$.

Now he could apply the condition that this is essentially zero. First, he cancelled the common multiplier e, getting $l - 2x - e$. Then he ignored the final term e, because e is arbitrarily small, getting $l - 2x$ equal to 0. This gives the answer $x = l/2$, as expected.

Although Fermat's conclusion is certainly correct, there is a logical problem with his argument. It is worth looking at it more closely, because it is historically significant. First, he divided by e, which is only allowed if e is not zero; however, at the final step, he took e to be zero. This difficulty is not fatal, but it took a very long time before mathematicians found a really satisfactory way to resolve it.

One small step . . .

Using algebra to study the effect of making small changes in varying quantities was to become the basis of a ground-breaking method of solving problems in mathematics. It is what we know as the differential calculus. In addition to his treatment of the maximum problem, Fermat had used a similar method when he solved the problem of finding tangents to the folium of Descartes. We now think of these as 'calculus problems', and Fermat takes pride of place among the precursors of the differential calculus.

The first person to develop the theory systematically was Isaac Newton, probably the greatest mathematician and scientist since Archimedes. Newton was born in the small village of Woolsthorpe in Lincolnshire in 1642, and entered the University of Cambridge in 1661. Although mathematics was not a major subject of study at that time, he acquired the latest books on the subject, and soon began to work on his own. In 1665–6 the University was closed because of an outbreak of the plague, and Newton returned to Lincolnshire. There, it is believed, he made some of his greatest discoveries.

Newton referred to variables, such as the coordinates (x, y) of a point moving along a curve, as flowing quantities. He began in the same way as Fermat, by making small changes to x and y. He supposed that o is a small quantity, and when x becomes $x + po$, then y becomes $y + qo$. The quantities p and q are what he called *fluxions*; they measure the rates at which x and y change, and their ratio

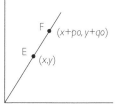

Figure 6.9 The change as the point E moves to F on the line $y = mx$.

q/p measures the relative rate of change of y with respect to x. Newton set out to calculate how these quantities depend on the nature of the curve at the point (x, y).

Suppose, for example, that the curve is a straight line with equation $y = mx$ as in Figure 6.9. The points $E = (x, y)$ and $F = (x + po, y + qo)$ both satisfy the equation of the line, and so we have the two conditions

$$y = mx, \qquad y + qo = m(x + po).$$

Here the algebra needed to find the ratio q/p is very easy. Subtracting the first condition from the second and cancelling o, it follows that $q = mp$, and so q/p is equal to m. In this case the relationship between p and q does not depend on the point (x, y): the ratio q/p is always equal to m, the slope of the line, and this is constant because the line is straight.

With this insight, we can use the same technique to investigate curves for which the slope is not constant, such as the circle $x^2 + y^2 = r^2$ (Figure 6.10). Here the algebra is slightly more complicated. After substituting (x, y) and $(x + po, y + qo)$ in the equation and subtracting, we obtain

$$(2xp + 2yq)o + (x^2p^2 + y^2q^2)o^2 = 0.$$

At this stage Newton's method was similar to Fermat's: divide by o and then ignore the last term, because it is a multiple of the small quantity o. The result is

$$xp + yq = 0, \quad \text{that is } q/p = -(x/y).$$

In this case q/p, the rate of change of y with respect to x, depends on the chosen point (x, y), as we might expect.

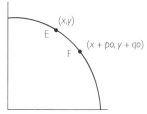

Figure 6.10 The change as the point E moves to F on the circle $x^2 + y^2 = r^2$.

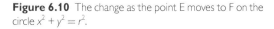

In geometrical terms, Newton's calculus of fluxions provided the complete solution to the problem of finding the tangent to a curve at a given point. By allowing o to approach zero we can make F as close as we please to E, so that the line EF approaches the tangent at E, the line that touches the curve at E. The ratio q/p of the fluxions is the slope of the tangent line, as required. Of course, the algebra can be nasty, but that did not deter Newton (his own examples were rather more complicated than those given here).

Newton was well aware that there was a problem about the status of the quantity o in his method. At the last stage in the calculation o is taken to be zero, but earlier it cannot be zero, because division by o is allowed. He made several attempts to explain away this difficulty, writing at length about what he called ultimate ratios, but he never quite settled the matter. With hindsight we can see that the underlying problem was the lack of a clear definition of the system of numbers that his symbols represented.

Another of Newton's great achievements was to discover a link between the method of fluxions and the problem of quadrature. His own account tells us that he originally discovered this link by looking at several examples, and noticing a striking pattern. As usual, his examples were rather complicated, but we can follow in his footsteps by looking at a simple case. The diagram (Figure 6.11) shows the area S bounded by the curve $y = x^2$ and the x-axis between the values $x = 0$ and $x = z$. We are looking for a formula for S in terms of z, which we shall regard as a variable. By using methods of quadrature like those of Archimedes and Fermat, Newton was able to work out that $S = z^3/3$. The key point, as Newton saw it, is that the rate of change of $z^3/3$, with respect to z, is z^2. Remarkably, the rule that defines the rate of change of the area is the same as the rule that defines the curve. Of course, Newton needed many similar examples before he could be sure that the relationship is not an accident. This far-reaching conclusion is now known as the Fundamental Theorem of Calculus. It tells us that the area bounded by any curve can be calculated by finding a rule whose fluxion is the rule that defines the curve.

An example will elucidate this amazing link between fluxions and areas. Suppose you require the area A bounded by the curve $y = x^4$, from 0 to z. You need only find the relationship between A and z for which the fluxion is z^4. Now, one of

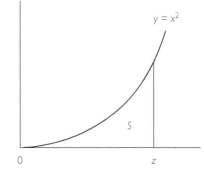

Figure 6.11 The area S can be found by fluxional methods.

the first rules of fluxional calculus, almost as important as the times-table in arithmetic, is that the fluxion of z^n is nz^{n-1}. Working backwards, you will find that the answer is $A = z^5/5$. This general principle is worth repeating yet again: the method of fluxions, applied in reverse, solves the problem of quadrature, just as it can be applied directly to solve the problem of tangents.

More about series

Newton made many fundamental discoveries in the 1660s, although none of them were published at that time.[7] The reasons for his reticence are complicated, and many books have been written about them. But this book is about mathematics, not mathematicians, and so we shall leave such speculation to those who aspire to be novelists or television pundits. We shall have more to say about Newton in Chapter 7, but there is one more of his early discoveries that belongs here: it is a remarkable extension of the binomial formula.

When Newton wrote about this work he began by explaining his notation for powers of a number b. Harriot (Figure 6.1) had written bb, bbb, $bbbb$ for the square, cube, and fourth power of b, but Newton preferred the now-standard b^2, b^3, b^4. When r (known as the exponent) is a whole number, b^r is defined simply as the product of r bs, and the fundamental property of exponents is therefore neatly expressed by the condition $b^r \times b^s = b^{r+s}$.

It is natural to ask whether the notation can be extended. What is the meaning of b^r when r is a negative number, or a fraction? For example, what is $b^{1/2}$? The answer is forced on us by the fundamental property, because we must have

$$b^{1/2} \times b^{1/2} = b^{1/2+1/2} = b^1 = b.$$

This means that $b^{1/2}$ is the number whose square is b; in other words, it is the square root of b. A similar argument shows that $b^{1/3}$ is the cube root of b, and $b^{1/n}$ is the nth root of b. Extending this idea to a general fraction m/n is straightforward: $b^{m/n}$ must be the mth power of $b^{1/n}$. For example, $8^{2/3} = (8^{1/3})^2 = 2^2 = 4$.

One of Newton's most useful discoveries was that the formula for the power of a binomial holds when the exponent is a fraction. To understand the significance of this result, it is helpful to look again at the situation when n is a positive whole number. We can think of the formula for $(a + b)^n$ as a description of what happens to the nth power of a when a is changed to $a + b$. In other words, it is an algorithm for calculating $(a + b)^n$ by adding successive terms to the first one, a^n. Explicitly $(a + b)^n$ is a sum of terms $T_0, T_1, T_2, \ldots, T_n$, where T_0 is a^n and

$$T_k = \binom{n}{k} a^{n-k} b^k.$$

This viewpoint suggests a good way of doing the calculations, because there is a neat relationship between each term and the next one:

$$T_k = \frac{n-k+1}{k} \times \frac{b}{a} \times T_{k-1}.$$

So the terms can be calculated successively, starting from $T_0 = a^n$. For example, suppose you know that $20^3 = 8000$, and you want to calculate 21^3. Writing 21^3 as $(20 + 1)^3$, and using the rule displayed above, the calculation goes as follows:

$$T_0 = 8000, \quad T_1 = \frac{3}{1} \times \frac{1}{20} \times 8000 = 1200,$$

$$T_2 = \frac{2}{2} \times \frac{1}{20} \times 1200 = 60, \quad T_3 = \frac{1}{3} \times \frac{1}{20} \times 60 = 1.$$

So you have worked out that $21^3 = 8000 + 1200 + 60 + 1 = 9261$.

When n is a whole number the calculation stops at the nth term T_n. Newton's great discovery was that we can use the same rules for expanding a binomial when the exponent is a fraction or a negative number instead of n. But in these cases the terms go on forever. In other words, we have an infinite series. Fortunately, as Archimedes, Harriot, Fermat, and Newton all knew, if the terms get smaller *in the right way*, then the infinite series will have a finite sum. Thus we have an ideal tool for finding approximations to numbers like the square root of 2 (that is, $2^{1/2}$) for which there is no exact answer.[8]

In the seventeenth century a new world of mathematics was created. It began with the deployment of algebraic symbolism to make arithmetic easier, and this remained a feature of the new mathematics, as exemplified by Newton's binomial series. But the use of algebra also transformed geometry, and led to the discovery of the calculus. Mathematics was now ready to play its part in some of the greatest achievements of humankind, such as explaining the motion of the moon, and actually going there.

CHAPTER 7

Mathematics Ascending

I f you were living in London at the end of the seventeenth century you might
have heard that someone called Newton was being employed by the government to look after the nation's coins. You might have assumed (correctly)
that his mathematical skills were useful for calculating such things as exchange
rates for foreign coins. But you would almost certainly be surprised by the great
advances in knowledge that were about to take place, partly as a result of Newton's
mathematical discoveries. Many of the new results would have appeared highly
theoretical at the time, but in due course they would find far-reaching practical
applications.

Calculus, coinage, and controversy

The method of fluxions, and other fundamental mathematical discoveries made
by Newton in the 1660s, remained unpublished for many years. The reasons are
complex, and seem to be personal, rather than mathematical. But whatever the
reasons, the facts are relevant to our theme, and must be recorded briefly.[1]

In his early researches Newton had made important discoveries in optics,
mechanics, and astronomy, as well as mathematics. He continued these researches
in the relative obscurity of Cambridge until, in 1687, he published his great work,
the *Mathematical Principles of Natural Philosophy*, usually referred to as the *Principia*, part of its Latin title. It contained a description of the physical world that
explains almost everything that we encounter in our daily lives, from the movement of the planets to the flight of a tennis ball. Although many of the results could
have been obtained by the method of fluxions, Newton chose to express most of
his arguments in geometrical terms, thus placing his book clearly in the classical
tradition. The *Principia* made Newton famous far beyond the narrow confines of
academe.

Within the academic world the fact that Newton had made great strides in
mathematics gradually became known. His discoveries had led to his appointment
as Lucasian Professor of Mathematics at Cambridge in 1669, and in 1671 he wrote

a long memoir describing some of his work on fluxions and infinite series. But he could not be persuaded to publish it. Nevertheless, his manuscript was seen by a few people, and in 1676 Henry Oldenburg, the secretary of the Royal Society, persuaded Newton to write two letters outlining some of his results to the German mathematician Gottfried Wilhelm Leibniz, of whom we shall say more shortly. These partial and unsatisfactory revelations of Newton's achievements were the root cause of a bitter and unedifying controversy.

In 1696 Newton's life took a new turn, when he was appointed to the position of Warden of the Tower Mint in London. Like so many aspects of his life, the reason for this change is not clear. It is sometimes thought that Newton's position at the Mint was a sinecure, and indeed his appointment in 1696 may have started in that way, because he retained his professorial chair in Cambridge. But he soon became involved in the problems of the English currency, which was at that time undergoing a major revision of the silver coins. In 1699 the post of Master of the Mint fell vacant. The Master was historically the 'master-worker', which implied a closer involvement in the day-to-day management of affairs, and Newton was appointed to the post. He continued as Master until his death in 1727, and throughout that period he had to wrestle with the problems created by a currency of coins made of precious metal.[2]

By this time, fixing the value of coins in circulation had become a far more complex problem than it had been in the thirteenth and fourteenth centuries, when Fibonacci and Pegolotti had been involved. Here is a short extract from one of Newton's many reports to the Treasury. It was written in 1701, and it concerns the valuation of the French coins that circulated in England at that time.

> The Lewis d'or passes for fourteen livres and the Ecu or French crown for three livres and sixteen sols. At which rate the Lewis d'or is worth 16s 7d sterling, supposing the Ecu worth 4s 6d as it is reckoned in the court of exchange and I have found it by some Assays. The proportion therefore between Gold and Silver is now become the same in France as it has been in Holland for some years.

Few people could make sense of all this. Those involved in the daily round of trade and commerce carried small money-scales and weights so that they could check the weight of the French gold Lewis d'or coins and the English gold guineas that circulated concurrently (Figure 7.1).

The guinea had been minted with a fixed weight and fineness since the 1660s, but its valuation varied, because the standard of value at that time was silver, and the relative price of gold and silver depended on economic circumstances. In the 1690s the value of the guinea had reached 28 shillings, but by 1701 it had settled around 21 shillings and 6 pence. Perhaps the most enduring legacy of Newton's time at the Mint was fixing the valuation of the guinea at 21 shillings in 1717. The guinea coin became obsolete about 100 years later, when gold became the official standard of value, and a new coin, the sovereign, was introduced. But for at least 150 years after that the word 'guinea', signifying 21 shillings, remained in use as the unit of account for fees charged by doctors and lawyers.

Figure 7.1 Money-scales and weights, c.1700. The weights are for checking the English gold guinea and half-guinea coins, and the French gold Lewis d'or and its half.

In Newton's time, the problem of establishing a relationship between the English and French coins was fraught with difficulty. The weight and fineness of the Lewis d'or differed from that of the guinea, and Newton's report, quoted above, was prompted by the former being altered. The government needed precise figures for such purposes as tax collection, and these entailed non-trivial calculations. Newton did many of the calculations himself, and his reputation gave authority to his recommendations.

Newton was assiduous in all his duties at the Mint. Not only did he insist on high standards of uniformity in the weight and fineness of the coins produced there, he also pursued counterfeiters relentlessly and took an interest in the production of commemorative medals. Despite this unremitting workload, he continued to work on mathematics and some (but not all) of his earlier work was eventually published. However, he had been overtaken by events elsewhere.

As already mentioned, in the 1670s Gottfried Leibniz had become interested in the two fundamental problems of the calculus—finding tangents to curves and calculating areas. His interest in these matters led to correspondence with Newton, via Henry Oldenburg, and eventually, in 1684, Leibniz began to publish his own work, which dealt with the subject in a new way. Many of the results in his

1684 paper had already been found by Newton, but they had not appeared in print. One very significant difference was that Leibniz's notation was much clearer than Newton's, although the printing techniques of the time did not always allow it to be displayed to the best advantage.

Leibniz denoted small changes in the variables x and y by dx and dy, and he called them *differentials*. He was able to work out the relationship between dx and dy when y depends on x in certain ways: for example, if $y = x^n$ where n is a positive integer, the rule is $dy = d(x^n) = nx^{n-1}\, dx$. He explained how to extend this rule to negative and fractional powers. (For a negative power, x^{-m} is defined to be $1/x^m$.) He also gave useful general rules, such as the rule that the differential of a product, $d(uv)$, is $udv + vdu$. In essence his methods were much the same as Newton's, but his differentials were easier to work with than Newton's fluxions, and his notation was soon adopted everywhere, except in England.

Leibniz also worked on the quadrature problem. His method was based on expressing the area as the sum of lots of small pieces, in the tradition of the ancient surveyors (and Archimedes, of course). This led him to suggest a new notation, an elongated S, for the sum of small contributions. We now call it the *integral* sign, $\int$. Leibniz was also familiar with Newton's 'Fundamental Theorem', which says that finding the area under a curve (the integral) and finding the slope of the tangent (the differential) are inverse problems. This principle is the basis for finding areas in terms of *indefinite integrals*, such as

$$\int x^n\, dx = \frac{x^{n+1}}{n+1}.$$

This rule is justified because the differential of the right-hand side is the expression following the integral sign. (This remains true if an arbitrary constant is added to the right-hand side.)

Leibniz's notation and rules were the foundations of what we still call the differential calculus and the integral calculus. They were easy to use, and the subject was taken up enthusiastically by many other mathematicians, leading to wide-ranging advances, some of which will be described in the following sections. However, we cannot leave Newton and Leibniz without referring to the bitter controversy that arose about the discovery of the calculus.[3] The facts are stated above but, in matters of this kind, facts are often less important than the opinions of important persons. From the historical point of view, it is fortunate that the controversy had little impact on the advance of mathematics.

Logarithms and exponents

One particular problem fitted awkwardly into the theory of Newton and Leibniz. The rule displayed above for finding the integral of x^n works not only when the exponent is positive, but also (when suitably interpreted) in most cases when the exponent is negative. But it fails dramatically when $n = -1$, because the right-hand side has zero as its denominator. So the general rule cannot be applied to the quadrature of the curve $y = x^{-1}$, that is, the hyperbola $y = 1/x$. This anomaly explains why Fermat's method of quadrature (Figure 6.7), which had been generally successful for curves of the form $y = 1/x^m$, had failed in the case $m = 1$. It was a serious problem, because the hyperbola was one of the famous curves discussed by the Greek geometers, and the quadrature of the hyperbola had therefore attracted much attention. Many interesting results were discovered by Gregoire de Saint Vincent in the period 1625–47, but his work was entirely geometrical, and consequently he missed the algebraic property that would have been his greatest discovery.[4] In the 1650s an Englishman, William Brouncker, used algebraic methods to calculate the area numerically, but he too did not state explicitly what we now regard as the key property. Although the details remain cloudy, it seems that by the late 1660s this property had become an accepted fact in the mathematical community.

As in Figure 7.2, let us denote by $H(a)$ the area bounded by the hyperbola $y = 1/x$, the x-axis, and the lines $x = 1$ and $x = a$. Then the most remarkable property of $H(a)$ can be expressed very simply in algebraic terms:

$$H(ab) = H(a) + H(b).$$

The significance of this result is that it has exactly the same form as the law of the logarithm, $\log(ab) = \log a + \log b$. For this reason $H(a)$ became known as the *hyperbolic* logarithm of a.

It seems that at first the hyperbolic logarithm was regarded merely as a curiosity, but the mathematicians of the day quickly realized that important questions needed to be answered. Can hyperbolic logarithms be calculated easily? Does the hyperbolic *anti*logarithm have any significance? What is the connection with the 'common' logarithms calculated by Briggs and others?

An answer to the first question was published by Nicolaus Mercator in 1668, in the form of an infinite series for calculating the hyperbolic logarithm. One small,

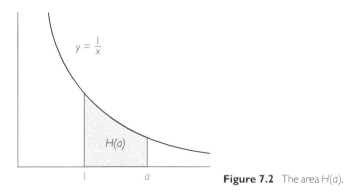

Figure 7.2 The area $H(a)$.

but significant, step was to write the variable a as $1 + x$, in which case the result comes out neatly as

$$H(1+x) = x - \frac{x^2}{2} + \frac{x^3}{3} - \frac{x^4}{4} + \frac{x^5}{5} - \dots \,.$$

Mercator's publication of this series alarmed Newton, because he had also discovered it. Probably this was what persuaded him to write an account of his work in 1669–1671 but, as we know, it remained unpublished. The memoir contained many relevant results, in particular the answer to another of our questions. After calculating the hyperbolic logarithm of 2 to many decimal places, and explaining how his method would work for any given number, Newton remarked that such calculations could be applied to the construction of tables of common logarithms, because the common logarithm of any number can be obtained by multiplying its hyperbolic logarithm by 0.4342944819032518, approximately.

Unlike some of his predecessors, Newton was not afraid of infinity, and he was happy to perform algebraic operations with infinite series. He used this method to study the problem of the hyperbolic antilogarithm: finding the number x such that the hyperbolic logarithm $H(1 + x)$ has a given value y. He boldly assumed that x could be written as an infinite series,

$$x = Ay + By^2 + Cy^3 + Dy^4 + \dots \,,$$

so that the problem is to find the numbers A, B, C, D, and so on. This can be done by substituting the series for x in the series for $H(1 + x)$. The algebra is tiresome, but the end justifies the means, because the series for $1 + x$, the number whose hyperbolic logarithm is y, turns out to be quite memorable: it is

$$1 + y + \frac{y^2}{2} + \frac{y^3}{6} + \frac{y^4}{24} + \dots \,.$$

The general term is simply y^n divided by $1 \times 2 \times 3 \times \dots \times n$. This remarkable series is usually known as the *exponential* series, for reasons that we shall now explain.

Modern notation makes the explanation clear, although it would have been less clear in the seventeenth century. Suppose that r and s are the hyperbolic logarithms of a and b. Then, as $H(ab) = H(a) + H(b)$, the hyperbolic logarithm of ab is $r + s$. Turning this statement around, the number whose hyperbolic logarithm is $r + s$ is ab, the product of the numbers whose hyperbolic logarithms are r and s. If, instead of the cumbersome phrase 'the number whose hyperbolic logarithm is y', we use the abbreviation $E(y)$, we obtain the rule

$$E(r + s) = E(r)E(s).$$

Now this is a familiar rule: it is just the rule for calculating exponents, $b^{r+s} = b^r b^s$, well known for many centuries, and expressed in symbolic form by Harriot,

Newton, and others. The value of $E(r)$ is simply the rth power of $E(1)$; this is a simple consequence of the exponential rule.[5]

So what is this number $E(1)$, the number whose hyperbolic logarithm is 1, and whose rth power is $E(r)$? It is clearly an important number, and its numerical value, 2.71828 ... is fairly easy to calculate from the infinite series. But in the seventeenth century it had to be referred to by the awkward phrase 'the number whose hyperbolic logarithm is 1'. When the delayed baptism of $E(1)$ eventually took place, it was an indirect result of the work of Leibniz. Soon after he began to publish his results on the calculus, the subject was taken up enthusiastically by two brothers, Jakob and Johann Bernoulli. These two quickly formulated most of what undergraduate students now learn about differential and integral calculus, and their rapid progress was the main reason why Leibniz's notation became standard, in preference to Newton's. As we shall see, several other members of the Bernoulli family also contributed to mathematics.

In the 1680s, while preparing his great work on the theory of probability, the *Ars Conjectandi*, Jakob Bernoulli came upon the exponential series. It seems that he rediscovered the work of Harriot on continuous compounding, which of course had never been published. Both Harriot and Bernoulli began with a binomial expansion, specifically a form of the finite series

$$\left(1 + \frac{y}{n}\right)^n = 1 + A_1 y + A_2 y^2 + A_3 y^3 + \ldots + A_n y^n.$$

Here the coefficients depend on the two occurrences of n on the left-hand side. When n is allowed to become large, Harriot and Bernoulli found that the coefficients approach the values 1, 1/2, 1/6, 1/24, and so on. So the result of allowing n to approach infinity is just the exponential series, as given above.

By the early years of the eighteenth century the exponential series was common knowledge, but its sum was still being described as 'the number whose hyperbolic logarithm is y'. For example, this term is used in a large tome entitled *Tables of Ancient Coins, Weights and Measures*, written by John Arbuthnot and published in 1727. The book is almost entirely concerned with scholarship of the antiquarian kind, but in the middle of a learned recital of the rates of interest charged in Roman times, Arbuthnot suddenly changes key.

> Monthly interest is higher than an annual one of the same rate, because it operates by compound interest. This suggests to me the following Problem. The rate per annum being given, to find the greatest Sum which is to be made of one Pound, supposing the interest payable every indivisible moment of time.

Arbuthnot proceeded to work out the answer to 'his' problem—as we know, it had previously been discussed by Harriot and Jakob Bernoulli. The 'indivisible moment of time' is a variable t, which can be arbitrarily small, and thus corresponds to $1/n$ in the discussion given above. Using essentially the same method, he worked out that 10000000 pounds invested for one year at 6 per cent should yield 10618364 pounds, approximately. This is over 18000 pounds more than the yield from a single annual payment.

In his calculation, Arbuthnot referred to 'the number whose hyperbolic logarithm is r', where r is the annual rate of interest. In our terminology, this is $E(r)$, the rth power of $E(1)$, the number whose hyperbolic logarithm is 1. When Arbuthnot's book was published, $E(1)$ still had no name, but it received one shortly afterwards, at the hands of the eighteenth century's finest mathematician, Leonhard Euler.

Euler's father had been a pupil of Jakob Bernoulli, and Leonhard himself studied with Johann Bernoulli before moving to St Petersburg in 1727. He remained there until 1741, then moved to Berlin until 1766, and then returned to St Petersburg until his death in 1783. He was extremely prolific, producing over 700 books and articles, as well as 13 children. It is thought that Euler first used the symbol 'e' for what we have called $E(1)$ in an unpublished letter written in 1728, the year after Arbuthnot's book was published. In 1731 he defined e explicitly as the number whose hyperbolic logarithm is 1, and it first occurs in print in the first volume of his *Mechanica* (1736). By introducing this notation, Euler dispelled the mists that had formed over many decades. It was now clear that the sum of the exponential series with parameter r is simply the rth power of e, that is, e^r.

Numbers, perfect and prime

The new theories of calculus and probability developed in the seventeenth century meant that mathematics could be used to illuminate many areas of natural science and human activity. At the same time some traditional parts of mathematics, such as the Theory of Numbers, were re-visited.

The study of the mathematical properties of whole numbers had begun in a blaze of glory with the proof, given in Euclid, that there are infinitely many primes. But the Greeks and their followers found the primes less interesting than the so-called Perfect Numbers, and for nearly 2000 years the primes seem to have been regarded as a curiosity. It was not until the seventeenth century that mathematicians began to notice that the primes have properties that are eminently worth studying; and it was not until the twentieth century that their properties were applied to real-life problems.

The Greeks had declared that a number is 'perfect' if it is equal to the sum of its factors, excluding itself. For example, 6 has the factors 1, 2, and 3, and 28 has the factors 1, 2, 4, 7, and 14, so these numbers are perfect because

$$6 = 1 + 2 + 3, \qquad 28 = 1 + 2 + 4 + 7 + 14.$$

Euclid gave a remarkable method of constructing such numbers, which can be roughly translated as follows.

> Suppose that, starting from a unit, numbers are taken in double proportion continually, until the whole added together is a prime number. Then the product of this whole multiplied by the last number is a perfect number.

In modern terms, this says that if $1 + 2 + 2^2 + \ldots + 2^k$ is a prime number p, then 2^k times p is a perfect number. When $k = 1$ and $k = 2$ this rule produces the numbers 6 and 28, and it also produces the next perfect numbers 496 and 8128. Euler proved that any even perfect number must be given by Euclid's rule, but to this day we do not know if there is an odd perfect number.

Throughout history the mathematical study of numbers has often been overshadowed by speculation about their mystical significance. That is what we call numerology, and its relationship with mathematics is similar to the relationship between astrology and astronomy. A typical numerologist of the late sixteenth century was Peter Bungus, who believed that the perfect numbers had special virtue, while certain other numbers like 666 were particularly beastly. In a work published in 1584, Bungus compiled a long list of numbers which he claimed were perfect, because he thought that the corresponding numbers $1 + 2 + 2^2 + \ldots + 2^k$ were prime. In fact many of Bungus's numbers were not perfect, as pointed out by Marin Mersenne in 1644.

Mersenne was an important figure in mathematical circles in the first half of the seventeenth century. He carried on an extensive correspondence with several leading mathematicians, including Descartes, Fermat, and Pascal, and helped to disseminate their results at a time when there were few mathematical journals. He also made significant contributions of his own. His approach to the study of numbers was essentially practical; he did not attempt proofs in the manner of Euclid, nor did he indulge in numerological speculation like Bungus. Mersenne's approach provided an important insight into the effectiveness of the algorithms of arithmetic. He knew that a sum of the form $1 + 2 + 2^2 + \ldots + 2^k$ is equal to $2^{k+1} - 1$, and he could easily have worked out the numerical value of 2^{k+1} by repeatedly multiplying by 2. This is a simple task, and some people even find it quite enjoyable. Mersenne did not need to do it himself, because tables of values of powers of 2 were available in textbooks of arithmetic, such as the one published by Nicholas Hunt in 1633 (Figure 7.3).

On the other hand, Mersenne observed that not all arithmetical procedures are so easy.

> It is one of the most difficult of all mathematical problems to exhibit a list of a collection of perfect numbers, just as also to understand whether given numbers consisting of 15 or 20 digits are prime or not, since not even a whole lifetime is sufficient to examine this, by any method so far known.[6]

This is a very significant remark. His point was that the only method available (at that time) for checking whether a number is prime or not was to try to find factors of it. But the algorithm for checking whether a given number has any factors is little better than brute force. A specific potential factor can be checked by long division, which is tiresome but possible. The problem is that there are too many potential factors: if m has 20 digits, then all numbers with up to 10 digits must be checked as potential factors of m. Certainly we can restrict the search to prime numbers,

Places or Numbers of progression.			
1	1	36028797018963968	56
2	2	18014398509481984	55
3	4	9007199254740992	54
4	8	4503599627370496	53
5	16	2251799813685248	52
6	32	1125899906842624	51
7	64	562949953421312	50
8	128	281474976710656	49
9	256	140737488355328	48
10	512	70368744177664	47
11	1024	35184372088832	46
12	2048	17592186044416	45
13	4096	8796093022208	44
14	8192	4398046511104	43
15	16384	2199023255552	42
16	32768	1099511627776	41
17	65536	549755813888	40
18	131072	274877906944	39
19	262144	137438953472	38
20	524288	68719476736	37
21	1048576	34359738368	36
22	2097152	17179869184	35
23	4194304	8589934592	34
24	8388608	4294967296	33
25	16777216	2147483648	32
26	33554432	1073741824	31
27	67108864	536870912	30
28	134217728	268435456	29

Figure 7.3 A table of powers of 2, from Nicholas Hunt's *Hand-maid to Arithmetick*, 1633. The number next to n is 2^{n-1}.

but then we must have a reliable list of all the primes with up to 10 digits before we begin. In fact there are hundreds of millions of them, and each one has to be tested by dividing it into *m*. That is why Mersenne said that not even a whole lifetime would suffice.

Mersenne concluded that this difficulty was the cause of the many errors of Peter Bungus. To show that a number is not prime, it is sufficient to find one factor, but to show that it is prime, we must show that there are no factors. Thus, except for the smallest values of *n,* it is very difficult to check whether a number of the form $2^n - 1$ is a prime. Of course that makes the problem more interesting. A list of powers of two, such as the one shown in Figure 7.3, suggests some obvious conjectures. For example, noting that the numbers

$$2^2 - 1 = 3,\ 2^5 - 1 = 31,\ 2^7 - 1 = 127,$$

are all primes, it is tempting to suggest that $2^n - 1$ is a prime whenever *n* is a prime. But this conjecture is easily shown to be false, because $2^{11} - 1 = 2047$ has the factors 23 and 89. Nevertheless, it is easy to show by elementary algebra that the converse result is true: if $2^n - 1$ is a prime then *n* itself must be a prime. So, the question for Mersenne and his contemporaries was: for which primes *p* is it true that $2^p - 1$ is a prime? Primes of the form $2^p - 1$ are now known as Mersenne Primes, and the search for them is still actively pursued.[7] Mersenne himself found several of them, the largest being $2^{19} - 1 = 524287$. He also claimed that $2^{67} - 1$ is prime, but here he was wrong, because

$$2^{67} - 1 = 193707721 \times 761838257287.$$

He had presumably tested lots of potential prime factors but, understandably, he had not got as far as 193707721. It was not until 1903 that the factors were found.[8] Back in Mersenne's time the prospect of an easy test for primality seemed hopeless, and indeed it remained so until the end of the twentieth century, when primality testing became important for practical reasons, and electronic computers allowed faster methods of calculation. Ironically, the modern methods depend ultimately on discoveries made in the seventeenth century by Mersenne's correspondent Fermat.

In 1640 Fermat wrote to Mersenne with three of his own results on the numbers $2^n - 1$. One of them was: 'when the exponent [n] is a prime number, the radical [$2^n - 1$] reduced by unity is measured by the double of the exponent'. In modern terms, Fermat claimed that if p is prime then $2^p - 2$ is a multiple of $2p$. This result implies that $2^{p-1} - 1$ is a multiple of p. For example

$$2^{7-1} - 1 = 63 = 7 \times 9, \quad 2^{17-1} - 1 = 65535 = 17 \times 3855.$$

Fermat did not give a proof in his letter, although he claimed to have found one. It was, he said, 'not without difficulty'. Probably the first proof was given by Leibniz, and it was based on his researches in combinatorics. In a manuscript (unpublished at the time) he observed that the binomial theorem provides an expression for 2^p as the sum of binomial coefficients, specifically

$$2^p = (1+1)^p = \binom{p}{0} + \binom{p}{1} + \binom{p}{2} + \ldots + \binom{p}{p}.$$

The first and last terms are both equal to 1, so $2^p - 2$ is the sum of the other terms, and it is easy to check, using Bhāskara's formula (Chapter 5), that each of them is a multiple of p. Hence $2^p - 2$ is a multiple of p and it follows (provided p is not 2) that $2^{p-1} - 1$ is also a multiple of p.

Fermat also noticed that the result can be extended to numbers other than 2. This is now known as Fermat's Little Theorem: if x is any one of the numbers $2, 3, \ldots, p - 1$, then $x^{p-1} - 1$ is a multiple of p. For example, when p is 7 we can make a table of values of $x^6 - 1$:

x:	2	3	4	5	6
$x^6 - 1$:	63	728	4095	15624	46655,

and check by division that they are all multiples of 7.

Nowadays Fermat's Little Theorem plays a very important part in our daily lives. As we shall explain in Chapter 11, it is one of the results behind the algorithms that are used to keep our electronic communications secure. However, Fermat's name is better known for another theorem. He stumbled upon it while reading the works of Diophantus, a mathematician in the Greek tradition of whom we know very little. All we know about him is that he lived for a time in Alexandria, in the period between about 100 BC and 350 AD, and he wrote several books on the arithmetical properties of whole numbers. These books were lost for a very long time, but some of them were rediscovered in the fifteenth century, and in his youth Fermat acquired a copy of a Latin translation by Claude Bachet, published in 1621. He studied it carefully. The second part contains a discussion of whole numbers that satisfy the Pythagorean Theorem, such as $3^2 + 4^2 = 5^2$ and $5^2 + 12^2 + 13^2$. While reading this, Fermat's mind strayed, and he wrote a note in the margin, in Latin.

> It is impossible for a cube to be a sum of two other cubes, or . . . in general for any number which is a power greater than the second to be written as a sum of like powers. I have a truly wonderful proof of this fact, which the margin is too narrow to contain.

This comment was a personal one, and it remained unknown during Fermat's lifetime. After his death in 1665 his son collected a number of similar marginal annotations, and published them in an edition of *Diophantus*, 'containing observations by P. de Fermat'.

The marginal note is what we know as Fermat's Last Theorem: the fact (and it is now an acknowledged fact) that there are no whole numbers x, y, z such that $x^n + y^n = z^n$, for any value of n greater than 2. Fermat's claim to have found a wonderful proof is unlikely, because for over 300 years mathematicians struggled to find one. They discovered many remarkable facts in the process, and built some marvellous mathematical structures, but the complete proof, for all values of n eluded them. It was not until the 1990s that Andrew Wiles made the final step. It was quite a moment.[9]

New kinds of numbers

When children are taught arithmetic they are told that 'you can't take 7 away from 4'. This makes good sense, especially if you are calculating with pebbles, and it was a long time before people were bold enough to go a step further. When mathematicians first discovered how to solve quadratic and cubic equations they had to consider several cases, because the coefficients were not allowed to be negative numbers. But gradually it became clear that working with a negative number like -3, the result of taking 7 from 4, was not only harmless, but actually very convenient. All the rules of arithmetic can be extended to the negative numbers, provided we are prepared to accept a few new rules, such as 'minus times minus is plus'.

Another new kind of number system was forced on the mathematical world as a logical consequence of its acceptance of negative numbers. A positive number like 25 has the square root 5; and if we allow negative numbers then −5 is also a square root of 25, because $(−5) \times (−5) = 25$. So 25 has two square roots, which is fine, but now there is another problem. What is the square root of −25? At first sight there is no such thing; indeed there appears to be no square root of any negative number, because the square of any positive or negative number is always positive. However, square roots of negative numbers had arisen in calculating the roots of quadratic and cubic equations, and mathematicians gradually realised that here is another opportunity for going boldly—in this case, by inventing a new kind of number. So $\sqrt{−25}$ had to be accepted into the realm of numbers. This was achieved by introducing a new symbol 'i' to stand for $\sqrt{−1}$, and saying that an expression of the form $a + ib$ is a *complex* number, where a and b are any numbers of the old kind, now said to be *real* numbers. Here too boldness was its own reward, because if the old rules of arithmetic are imposed, with one new rule $i^2 = −1$, everything goes as it should. For example, it is easy to check that −25 has two square roots, 5i and −5i. At one time it was customary to use the words 'imaginary number' rather than 'complex number', but that is an unfortunate misnomer. If we accept −1 as a number, there is no reason to dismiss its square root i.

Complex numbers are not an academic curiosity. Their crucial importance in mathematics was first shown by Euler in the 1740s. He was an enthusiastic user of infinite series, sometimes in a manner that borders on the reckless, and one important discovery that resulted from his enthusiasm was that the complex numbers provide a link between apparently unrelated parts of mathematics. In his *Introductio in Analysin Infinitorum* (1748), he described his own approach to the exponential and logarithmic series. As we have seen, this amounted to showing that the powers of e, the number 2.71828 . . . , are given by the series

$$e^x = 1 + x + \frac{x^2}{2} + \frac{x^3}{6} + \frac{x^4}{24} + \frac{x^5}{120} + \dots .$$

Euler's great discovery was based on what happens when x is replaced by ix in this formula. Using only the rule that $i^2 = −1$, he was able to write the result neatly in the form of a complex number $a + ib$:

$$e^{ix} = \left(1 - \frac{x^2}{2} + \frac{x^4}{24} + \dots\right) + i\left(x - \frac{x^3}{6} + \frac{x^5}{120} + \dots\right).$$

Euler noticed a link between the two parts of this series and the calculations involved in the measurement of triangles, as used by navigators and surveyors. Using only simple algebra, he found infinite series for what we now call the trigonometric functions, *sine* and *cosine*, denoted by $\sin x$ and $\cos x$. Remarkably, these series are just the parts a and b of the series for e^{ix} and so, with a mighty bound, he concluded that

$$e^{ix} = \cos x + i\sin x.$$

Thus Euler had demonstrated that the exponential function and the trigonometric functions are linked in a very simple way, in the realm of complex numbers. After this, there could be no doubt that complex numbers were destined to play an important part in mathematics, and indeed they found many practical applications, for example in electrical engineering.

One special case of Euler's result is especially noteworthy. Putting x equal to π leads to the formula

$$e^{i\pi} = -1.$$

Here we have the amazing fact that two fundamental constants π and e are linked in a very simple way, if we look at them in the right context. There is nothing supernatural about this link; it is an inevitable consequence of our natural conceptions of arithmetic and geometry.

All kinds of wonderful things

Many fine mathematicians followed in Euler's footsteps. The greatest of them was Carl Friedrich Gauss (1777–1855), who expanded the frontiers of the subject to the point where, even today, few people can comprehend all his achievements. His prodigious mathematical talents were recognized at a very early age. In the early 1790s, while under the patronage of the Duke of Brunswick, he was given a book of tables of logarithms which also contained a long list of prime numbers. He was presumably aware of Euclid's proof that the list goes on for ever, but he noticed that it seems to thin out as the size increases. So he tried to discover a rule that would describe this behaviour. Here the tables of logarithms came in useful, because he noticed that the number of primes less than any given number n is approximately $n/\log n$ (where the logarithm is hyperbolic). He did not publish this result, presumably because he could not prove it, but in 1798 a modified form of the same conjecture was published by Adrien-Marie Legendre, one of a group of French mathematicians who, like Gauss, had been inspired by the work of Euler. Gauss himself later discovered a slightly better approximation, but the proof eluded him, and the Prime Number Theorem (as it is now known) remained a conjecture for nearly a century.

In 1801 Gauss published his *Disquisitiones arithmeticae*, an exposition of the known results on whole numbers, together with many new results of his own. He began by introducing some new terminology. He proposed that, if $a - b$ is a multiple of m, then we should say that

a is congruent to b with modulus m.

When the modulus m is 5 we find, for example, that 6 is congruent to 1, 24 is congruent to 4, and −2 is congruent to 3; every number, positive or negative, is congruent to one of the numbers 0, 1, 2, 3, 4. In the language of the ancients, this number is just what is left over when the given number is divided by 5.

The modern point of view is that the congruence relation defines a new kind of arithmetic, which we call *modular arithmetic*, or 'mod *m* arithmetic'. The crucial step is to think of the class of all numbers that are congruent with respect to a given modulus as a new kind of number. For example, in mod 5 arithmetic there are five classes, which we denote by the symbols 0, 1, 2, 3, 4. The rules of ordinary arithmetic imply that the relation of congruence is well-behaved with respect to addition and multiplication: thus if a_1 and b_1 are respectively congruent to a_2 and b_2, then $a_1 + b_1$ is congruent to $a_2 + b_2$ and $a_1 \times b_1$ is congruent to $a_2 \times b_2$. It follows that the classes 0, 1, 2, 3, 4 can be added and multiplied by making simple modifications to the ordinary rules, as shown in the tables (Figure 7.4).

The *Disquisitiones* of Gauss contain many fundamental results on modular arithmetic, certainly enough to show that it is not just a matter of abstraction for its own sake. In particular, when the modulus is a prime number *p* it turns out that mod *p* arithmetic has some remarkably beautiful properties. But beauty is not their only attribute; they are so useful that we could not now live our lives without them, as we shall see in Chapter 11.

One thing to remember is that these results hold for any prime number *p*, however large. We tend to think in terms of small primes, like 5 and 113, but the theorems hold for all primes, such as

$$p = 1907180854589209641162363757488357797106749590673031653701683922$$
$$6001220767984427385832966637999862924555166101.$$

But this *p* is really quite small too: it has only 108 digits, whereas the largest number currently known to be prime is the Mersenne prime $2^{57885101} - 1$, which has over 17000000 digits.

Translated into the language of modular arithmetic, Fermat's Little Theorem says that, in mod *p* arithmetic, every non-zero element *x* is such that x^{p-1} is congruent to 1. Euler noticed that there is always some element *r* for which the powers $r, r^2, r^3, \ldots, r^{p-1}$, are all different, and as there are $p-1$ of them, these powers are all the non-zero elements. As far as we know, Euler did not have a complete proof of this fact, but soon after his death it was proved by Legendre, and by Gauss in his

ADD	0	1	2	3	4		MULTIPLY	0	1	2	3	4
0	0	1	2	3	4		0	0	0	0	0	0
1	1	2	3	4	0		1	0	1	2	3	4
2	2	3	4	0	1		2	0	2	4	1	3
3	3	4	0	1	2		3	0	3	1	4	2
4	4	0	1	2	3		4	0	4	3	2	1

Figure 7.4 Addition and multiplication tables for mod 5 arithmetic.

Disquisitiones. Any *r* with this property is called a *primitive root* of *p*: for example 2 is a primitive root of 5, because in mod 5 arithmetic $2^1 = 2$, $2^2 = 4$, $2^3 = 3$, $2^4 = 1$. One consequence of this result is that in mod *p* arithmetic, every number (except 0) has an inverse; for every *x* there is a *y*, such that the product of *x* and *y* is 1. Another useful consequence is that 'logarithms' can be introduced.[10] Given all these remarkable properties, it should come as no surprise that the wonderful world of modular arithmetic is still being explored in the twenty-first century.

Gauss had many scientific interests, which led him to make advances in almost all branches of mathematics. For example, his interest in astronomy, sparked by the discovery of the asteroids Ceres and Pallas, encouraged him to invent methods of calculating orbits from a small number of observations. Other astronomical problems inspired him to make important contributions to statistics, and these will be described later. His work on electricity and magnetism led to the invention of a form of electric telegraph, and a unit of magnetic flux density is named after him.

Much of Gauss's mathematical work remained unpublished in his lifetime, partly because he was always looking for better methods of proof. His work on geometry led him to consider the status of Euclid's axiom that there is exactly one line passing through a given point that is parallel to a given line. There had been many attempts to prove that this statement can be deduced from Euclid's other axioms, but Gauss concluded that such attempts were futile, and around 1830 his view was verified independently by János Bolyai (1802–60) and Nikolai Lobachevskii (1793–1856).

Their method was to show that systems of *non-Euclidean* geometry are logically consistent. It is possible to define geometries in which there are many lines through a given point that are parallel to a given line—or indeed no such line. Nowadays our picture of the real world is based on geometrical ideas far broader than those allowed by Euclid, and such possibilities are unsurprising. But in Gauss's time many people believed that Euclidean geometry was a law of nature, expressing a basic truth about the real world, and for them the possibility of non-Euclidean geometry was a great shock.

To the limit, carefully

By the beginning of the nineteenth century the calculus of Newton and Leibniz had found many applications, and it was being used without qualms about its foundations. But persons of a nervous disposition (mathematicians) were still worried. One of the most pungent criticisms had been expressed by George Berkeley, an Irish bishop, in his book *The Analyst, or a Discourse Addressed to an Infidel Mathematician* (1734). He wrote of 'the ghosts of departed quantities', referring to the small increments that were initially not zero, but were finally put equal to zero to obtain the required answers. Such criticisms were troubling. A related problem was to define precisely the number-system in which the calculus operates. Whole numbers and fractions were beyond doubt, and the use of decimal notation seemed to cope with the other numbers that are needed, like $\sqrt{2}=1.41421\ldots$, $\pi=3.14159\ldots$, and e $=2.71828\ldots$. But, in fact, the decimal notation is just a way of writing these numbers as infinite series of fractions; for example $1.41421\ldots$ is another way of writing

$$1+\frac{4}{10}+\frac{1}{100}+\frac{4}{1000}+\frac{2}{10000}+\frac{1}{100000}+\ldots.$$

So the problem remains: to explain how an infinite number of terms can have a finite sum. The crucial step is to clarify the notion of a *limit*. For example, if the dependence of y on x is given, and consequently the relationship between the differentials dy and dx is known, how should the limit of the ratio dy/dx be defined? As dx approaches 0 so does dy, and we must avoid the trap of trying to assign a meaning to $0/0$. Gradually there emerged a form of words that is logically precise. We must say that the limit is L if we can make dy/dx as close as we please to L by taking dx to be sufficiently close to 0. Similarly, for the sum S of an infinite series we must insist that we can make the sum S_n of n terms as close as we please to S by taking n to be sufficiently large.

In the eighteenth century several mathematicians had attempted to frame proper definitions of these basic notions, but they all relied to some extent on intuitive assumptions. The need for truly rigorous definitions was stressed by Bernard Bolzano (1781–1848). In 1817 he published a short pamphlet with a title that can be roughly translated as 'Purely analytic proof of the theorem that between two values which give results of opposite sign there lies at least one root of the equation'. This result was one that had long been taken for granted when finding roots of equations. For example, to locate a root of the equation $x^3+x-1=0$ we might begin by calculating some values:

$$
\begin{array}{lccccc}
x: & -2 & -1 & 0 & 1 & 2 \\
x^3+x-1: & -11 & -3 & -1 & 1 & 9.
\end{array}
$$

On the basis of this table it is intuitively obvious that the equation has a root between 0 and 1. The value of x^3+x-1 is negative when x is 0 and positive when x is 1, and so its value must be 0 at some intermediate point. Of course, this argument depends on the intuition that the values of x^3+x-1 change continuously, as we assume when we draw a picture of the curve $y=x^3+x-1$.

Bolzano wanted to avoid intuition, and he succeeded in giving a definition of continuity from which the result can be derived. A few years later, possibly having read Bolzano, Augustin-Louis Cauchy used similar definitions as the starting point for his influential *Cours d'Analyse*.

The work of Bolzano and Cauchy clarified the basic notions of limits and continuity, but one problem remained. Their proofs assumed that the number-system has no gaps. This too seems to be intuitively obvious, but at that time there was no logical justification for it. Although it was possible to say precisely what is meant by a limit, it was not possible to guarantee that the limit exists. The difficulty was finally overcome by Richard Dedekind (1831–1916) and Georg Cantor (1845–1918). They gave explicit constructions of a number-system which has no gaps: the numbers that we call *real*.

By the end of the 1870s the foundations appeared to be sound, because the real number-system accurately represents our intuitive picture of the points on a straight line. Nowadays we can teach the foundations of mathematical analysis to undergraduates without worrying about 'the ghosts of departed quantities'. But that does not mean that mathematics is an open book. At the start of the twentieth century mathematicians began to find that the system has some counter-intuitive properties: for example, when we try to measure an arbitrary set of real numbers.[11] This particular difficulty has profound implications for our understanding of the notion of probability, and its application to practical problems. However, it would be wrong to end this account of modern analysis on a pessimistic note. The subject is not only sound, but also powerful. The Prime Number Theorem conjectured by Gauss was proved at the end of the nineteenth century by Jacques Hadamard and Charles de la Vallée Poussin, using methods based on complex numbers. And at the end of the twentieth century, Andrew Wiles' proof of Fermat's Last Theorem employed a formidable barrage of techniques from many branches of mathematics, so that no margin is wide enough to contain even a summary of it.

CHAPTER 8

Taking a Chance

Y ou win some, you lose some. When you toss a coin, the outcome (heads or tails) is uncertain. But when you repeat the experiment many times, you expect the numbers of heads and tails to be about the same. This chapter describes how a satisfactory theory of such events was developed, and how it can help you solve problems such as working out how much to expect from your financial investments.

Great expectations

A popular game in ancient times involved throwing several small pieces of bone, and betting on the way in which they fell.[1] In due course it became common to carve the bone into a cube, and to put symbols representing the numbers 1, 2, 3, 4, 5, 6 on its six faces. We shall call this a *dice*, following the common practice of using the same word for one or several of these objects, although technically 'dice' is a plural form. From our point of view the important point is that the symmetrical shape of the cube justifies the assumption that the dice is fair, which means that when it lands the number shown on its uppermost face is equally likely to be any one of the numbers 1 to 6. (This apparently simple remark is loaded with philosophical difficulties, but we shall set them aside for the time being.) Historically, the assumption of fairness was the justification for applying mathematical methods to the study of dice-games.

Suppose that three fair dice are thrown, and the players bet on the sum of the three numbers that turn up. The possible totals range from 3, when all three dice show 1, to 18 when all three dice show 6. People who played this game regularly would have noticed that these particular totals happen infrequently, and they would bet accordingly. They might also have guessed that a total like 11 turns up more frequently than 3 or 18 because it can occur in several different ways (Figure 8.1).

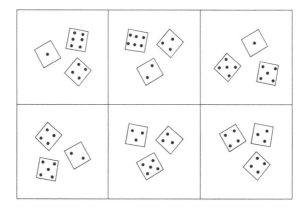

Figure 8.1 The six partitions of 11 that can result from three dice.

Mathematically speaking, we have here the six *partitions* of 11 into three parts, subject to the condition that each part is one of the numbers 1 to 6. For each of the possible totals it is easy to work out the corresponding number of partitions: there are 56 in all.

Total:	3	4	5	6	7	8	9	10	11	12	13	14	15	16	17	18
Partitions:	1	1	2	3	4	5	6	6	6	6	5	4	3	2	1	1

To analyze the game completely, a little more thought is needed, because a partition may occur in more than one way. Each dice can show six possible values, and when the three dice are thrown at the same time there are $6 \times 6 \times 6 = 216$ possible outcomes. Suppose we distinguish the dice, for example by calling them red, green, and blue. The partition with the numbers 5, 4, 2 occurs when these three numbers are shown on the three dice in any order; it does not matter which is on the red dice, which is on the green dice, and which is on the blue dice. In other words, the partition $5 + 4 + 2$ can occur in six ways, corresponding to the $3 \times 2 \times 1$ permutations of the numbers 5, 4, 2. The same is true for any partition composed of distinct numbers. However, a partition like $5 + 5 + 1$ can occur in only three ways: 1 can be shown on any one of the three dice, and the other two dice must both show 5. Referring to the diagram, we see that three of the partitions of 11 occur in six ways, and the other three in three ways, so the total number of occurrences of 11 is 27, out of the 216 possible outcomes.

An analysis of the game with three dice is contained in a long narrative poem written in Latin around 1250. In this poem the relationship between the 56 partitions and the 216 possible outcomes is described very clearly. Among the many manuscript copies of this poem, entitled *De Vetula*,[2] the one in the British Library is annotated with drawings and a table giving the relevant numbers for each total.

Totals		Partitions	Outcomes
3	18	1	1
4	17	1	3
5	16	2	6
6	15	3	10
7	14	4	15
8	13	5	21
9	12	6	25
10	11	6	27
In all:		56	216

Because the analysis in *De Vetula* is unusual for its time and place, some historians have suggested an Islamic origin for the calculations. There is indeed a resemblance to the way in which the relative frequency of letters in a coded message was used by the Islamic cryptographers.

One reason why so many copies of *De Vetula* have survived may be that it offered a sure way of making a profit from gambling. Those who were lucky enough to obtain a copy would have kept it safe. The typical medieval gambler had no inkling of a rational explanation of how the dice might fall: in fact, there were many who considered that attempting to make a prediction of any kind was trespassing into the realm of the deities. The tangled web of vague ideas about risk and luck that surrounded the so-called Games of Chance was part of their attraction. In this context, we can imagine a scenario in which a medieval entrepreneur (nowadays known as a bookmaker) has set up a game in which players are invited to bet on the outcome of a throw of three dice. A player who bets correctly will receive one lira from the bookmaker. If the game is fair, the bookmaker should ask for stakes in line with the numbers in *De Vetula*: for example, a player betting on 11 has 27 chances out of 216, and should put up a stake of 27/216 of a lira (30 denari). Clearly the bookmaker can make a profit simply by asking for stakes that are higher than the fair ones: some throws will result in a loss for him, but when the dice are thrown repeatedly the stakes will exceed the payouts. In the twenty-first century this principle is the foundation for a worldwide billion-dollar industry.

The medieval interest in the mathematics of gambling was partly driven by analogies between dice-games and the problems of finance and commerce. By the fourteenth century Italian merchants had begun to formulate contracts covering such matters as marine insurance. This involved estimating the risk incurred when a shipload of merchandise is sent on a voyage and, as in gambling, the risk had to be measured in a way that can be expressed in monetary terms. There was a vague feeling that the stake (insurance premium) should be determined by the expectation of gain or loss, but a mathematical basis for this notion was not yet available.

The development of a sound foundation was the result of work on a problem that resembled the problems of the market-place. It is known as the Problem of

Points. The scenario is similar to a 'set' of tennis, where two people play games repeatedly until one player reaches a certain number of successes. The players put up equal stakes, and the player who wins the set takes the lot. (For simplicity, we ignore the rules about tie-breaks.) The problem is this: if the set is interrupted before the end is reached, how should the stakes be divided?

There is a manuscript dating from around 1400, probably intended for the pupils in an Italian abacus school, which contains the correct solution to the case when three successes constitute a win and the play is interrupted with the score at 2–0. In this case the stakes should be divided in the ratio 7:1. However, the same manuscript also deals with another case, for which a wrong answer is obtained. Clearly there was no sound foundation for a general solution at that time, and it was to be a long time before one was found. Luca Pacioli discussed the problem in his *Summa*, and he suggested that the stakes should be divided in the same ratio as the current score: if the score is 4–3 when play is interrupted, the stakes should be divided in the ratio 4:3. Unfortunately Pacioli's rule can lead to unreasonable results. For example, if play is interrupted with the score at 1–0, the second player will receive nothing, which is clearly not a reasonable assessment of that player's expectation.

A rational solution to the problem of points did not emerge until the seventeenth century. It was based on a definition of *probability* that enabled *expectation* to be defined in a consistent way. The details were worked out in correspondence between Fermat and his friend Blaise Pascal, beginning in 1654. We shall explain Fermat's method by considering the case when seven successes are needed to win and the play is interrupted with the score at 5–4. Fermat began by suggesting that the prospects of the two players, Alice (A) and Bob (B), should depend only on the number of successes in the unplayed games that they would need to win. If we call these numbers a and b, then $a = 2$ and $b = 3$ in our example. Exactly the same values would apply if a total of 8 were required with the score at 6–5. Fermat also made the significant observation that one player must win after at most $a + b - 1$ additional games. This is because if Alice has won less than a games, and Bob has won less than b games, then they can have played at most $(a - 1) + (b - 1) = a + b - 2$ games. So in the case when $a = 2$ and $b = 3$, at most four more games are needed.

It is easy to make a list of the possible outcomes of these four games: there are $2 \times 2 \times 2 \times 2 = 16$ possibilities in all.

<u>AAAA</u>	<u>AAAB</u>	<u>AABA</u>	<u>AABB</u>	<u>ABAA</u>	<u>ABAB</u>	<u>ABBA</u>	ABBB
BBBB	BBBA	BBAB	<u>BBAA</u>	BABB	<u>BABA</u>	<u>BAAB</u>	<u>BAAA</u>.

Examining the list we see that Alice will win in all the cases underlined; so she has 11 chances and Bob has only 5. If we assume that each of the 16 possibilities is equally likely, then the stakes should be divided in the ratio 11:5.

Some debatable points in Fermat's argument were thrashed out in his correspondence with Pascal.[3] The fact that the four additional games are not always needed raises some questions, but Pascal was able to show that Fermat's conclusion was nevertheless correct. In the course of his analysis he was led to study the combination numbers $\binom{n}{r}$ which, as we know (Chapter 5), are simply the binomial coefficients. The outcome of Pascal's work was a classic book, *The Arithmetical Triangle*, in which he gave the first full exposition in modern terms of the triangular pattern formed by these numbers. Armed with this information he was able to find the general solution of the Problem of Points, without having to list all the possibilities. The key observation is that Alice wins in all cases when Bob does not achieve b or more successes in the $a + b - 1$ unplayed games. In our example, the possibilities in which Bob has 0, 1, or 2 successes in the four unplayed games result in a win for Alice, and the number of possibilities is therefore

$$\binom{4}{0} + \binom{4}{1} + \binom{4}{2} = 1 + 4 + 6 = 11.$$

Similarly, a win for Bob occurs in the cases where Bob is successful in 3 or 4 games, and the number of possibilities is

$$\binom{4}{3} + \binom{4}{4} = 4 + 1 = 5.$$

The general notion that emerged from the Fermat-Pascal analysis is that of assigning to an event a number p that measures its *probability*. The simplest situation is when we are given n events, which do not overlap, and which exhaust all the possible outcomes. Then the probabilities of these events, denoted by $p_1, p_2, \ldots, p_n$, are numbers between 0 and 1, and they must satisfy the condition

$$p_1 + p_2 + \ldots + p_n = 1.$$

In our example on the Problem of Points the elementary events were the 16 possible outcomes of the four unplayed games, as listed above. Each one was assumed to be equally likely, so there are sixteen probabilities $p_1, p_2, \ldots, p_{16}$, each of them equal to 1/16. The probability of a compound event, such as 'A wins,' is obtained by adding the probabilities of the 11 events that comprise it, so we have

$$Probability(\text{A wins}) = \frac{1}{16} + \frac{1}{16} + \ldots + \frac{1}{16} = \frac{11}{16}.$$

It turns out that this simple set-up is an adequate foundation for describing many situations in which there is an element of chance. Using it to formulate a definition of *expectation* produces a theory that clears up many problems that had previously appeared mysterious. In our example, suppose both Alice and Bob have staked 1 lira; then if Alice wins she will gain 1 lira and if she loses she will gain −1 lira. So Alice's expectation is

$$Probability(\text{A wins}) \times 1 + Probability(\text{A loses}) \times (-1) = \frac{11}{16} - \frac{5}{16} = \frac{3}{8}.$$

Generally, the expectation of gain is obtained by adding the terms $p \times g$, where p is the probability of gaining g.

Like Newton's method of fluxions, the seventeenth-century theory of probability was not quite the last word on the subject. In the rest of this chapter we shall explain how the theory developed, and was used to make models of more complicated situations.

Games and strategies

The basic rules for calculating probabilities and expectations, as described by Fermat and Pascal, were soon applied to gambling problems of many kinds. Jakob Bernoulli worked on this subject from about 1684 until he died in 1705, but his book, entitled *Ars Conjectandi* (The Art of Conjecturing), was not published until 1713. By then Pierre de Montmort had published the first version of his *Essai d'Analyse sur les Jeux de Hasard* (Essay on the Analysis of Games of Chance) in 1708, and a much enlarged version appeared in 1713. An illustration from Montmort's book shows several games being played in what appear to be very splendid surroundings, possibly one of Louis XIV's chateaux. Among the people depicted are Montmort himself, and (it is thought) Minerva, the goddess of wisdom.

One of the card games played by the French grandees was known as Le Her. Although the game itself is simple, its analysis required extensive calculations. The details were worked out in letters between Montmort and Nikolaus Bernoulli[4] and their results were published in the second edition of Montmort's *Essai*. The special significance of this work is that it led to the first known example of an entirely new way of looking at games.

Le Her is a card game for two players, whom we shall call Min (Minerva) and Mon (Montmort). The standard pack of 52 cards is used, with the cards valued in the ascending order 1, 2, 3, 4, 5, 6, 7, 8, 9, 10, Jack, Queen, King, the suits being irrelevant. The game has four stages, as follows.

1. One card from the pack is dealt to each player. If Mon has a King, he wins and the game is over, otherwise the game continues.
2. Min may either hold her card, or exchange it with Mon's card.
3. Mon may either hold his card or exchange it with a random one from the pack.
4. If Min has the higher card she wins, otherwise Mon wins (this includes the case when the cards are the same).

Note that if Min has opted to exchange at stage (2), both players will know both cards, and the option for Mon at stage (3) is determined: he must hold if he knows he will win, and exchange if not. He has a real choice at stage (3) only when Min has opted to hold at stage (2).

The analysis of the game depends on the strategy employed by each player. For each of the 13 card-denominations, a player may opt to hold or exchange when holding that card, which means that there are, at first sight, $2 \times 2 \times \ldots \times 2 = 2^{13}$ possible strategies. But almost all of them can be set aside. Let us denote by Hc the strategy:

hold with card c or higher, otherwise exchange.

Montmort and Bernoulli had worked out that there are only two strategies that are worth considering for each player, because all the others can be shown to be inferior to them. For Min the two strategies are H7 and H8, and for Mon they are H8 and H9. Suppose that Min plays strategy H7 and Mon plays H8. Montmort and Bernoulli carried out a long and complicated series of calculations, based on the probability of the cards dealt at stage (1), and the probability that a specific card is drawn from the pack at stage (3). They concluded that in this case Min can expect to win 2828 out of 5525 games. Similar calculations for the other cases led to the numbers shown in the table below.

	Mon plays H8	Mon plays H9
Min plays H7	2828	2838
Min plays H8	2834	2828

Thus far the analysis is standard, although complicated in detail. The main conclusion is that Min has a slight advantage, as she is bound to win at least 2828 games, which is rather more than one-half of 5525. However, it appears that this is the best she can do, because if she plays H7, Mon will play H8, and if she plays H8, Mon will play H9. The question is: how can she make use of the slightly better results offered by the other entries in the table?

The answer to this question was conveyed in a letter to Montmort from a certain Monsieur de Waldegrave, dated 13 November 1713. Waldegrave is a shadowy figure,[5] but his suggestion was clear: Min should mix her strategies. Specifically, at each play she should choose either H7 or H8 according to a random mechanism. She could have a bag containing black tokens and white tokens, draw one token every time, and play H7 if it is black, H8 if it is white. In mathematical terms this corresponds to playing H7 with probability x and H8 with probability $1 - x$, where x is determined by the ratio of black tokens to white tokens. Nowadays we call this the *mixed strategy* $(x, 1 - x)$.

To see why the mixed strategy works, we can simplify the situation by considering only how the numbers in the table improve on the baseline 2828. We shall refer to these numbers as the *payoffs*:

$$\begin{matrix} 0 & 10 \\ 6 & 0 \end{matrix}.$$

Suppose Min plays the mixed strategy $(x, 1 - x)$. If Mon plays H8, she expects a payoff 0 with probability x and a payoff 6 with probability $1 - x$. In other words,

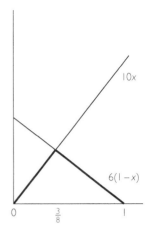

Figure 8.2 Min's expectations for the mixed strategy $(x, 1-x)$ against Mon's strategies H8 and H9.

her expected payoff is $6(1 - x)$. If Mon plays H9, her expected payoff is 10 with probability x and 0 with probability $1 - x$, that is, $10x$. So Min is guaranteed a payoff equal to the smaller of these quantities, irrespective of which strategy Mon chooses. At this point, a simple diagram helps (Figure 8.2).

The bold lines in the diagram represent the smaller of $6(1 - x)$ and $10x$ for each value of x. This is the guaranteed minimum payoff for Min, irrespective of the strategy adopted by Mon. So clearly Min should choose the value of x in such a way that this guaranteed minimum payoff is as large as possible: in other words, she should maximize her minimum. It is clear from the diagram that the best choice of x is where $6(1 - x) = 10x$. This occurs when $x = 3/8$, so Min's payoff from this particular strategy is $10 \times 3/8$, or three and three-quarter extra games won.

These results were obtained by Waldegrave in 1713. He did not give the details of his analysis, but he knew that if Min plays H7 three times out of eight and H8 five times out of eight, she can improve on the 2828 out of 5525 games that she can expect to win by using only one of the pure strategies. In fact, she can expect to win an extra three and three-quarter games.

Underlying Waldegrave's discovery there is a very remarkable piece of mathematics, which was not fully appreciated until the twentieth century, when mathematicians began to study what we now call the Theory of Games. Suppose we carry out a similar analysis, but from Mon's point of view rather than Min's. Then we find that he too can improve on his pure strategies by using a mixed strategy. His aim must be to minimize his maximum loss, and this can be done in the same way as Min maximized her minimum gain. The remarkable feature is that the outcome for Mon turns out to be the reverse of that for Min: his smallest possible loss is the same as her greatest possible gain.

The first attempt to apply modern mathematical methods to games of strategy was a paper written by Ernst Zermelo in 1913. He discussed some basic questions about the possible outcomes of the game of chess, but his results were very general—and they are frequently misquoted.[6] The detailed analysis of chess is complicated, but the game of Le Her is much simpler because there are essentially just two pure strategies for each player. Nowadays it is called a *two-person zero-sum* game, with finitely many pure strategies. The name indicates that the payoff for one player is exactly the opposite of the payoff for the other player: whatever one wins, the other loses. The payoff table has a finite number of rows and columns, and its properties can be studied using analytical techniques that are now the everyday tools of mathematics. The modern treatment of this kind of game was initiated by Emile Borel, in a series of papers written between 1921 and 1927. He found that in many cases the solution has the remarkable property referred to in the previous paragraph, and he conjectured that this result was true in general. In 1928 John von Neumann gave a complete proof of this fact: for any two-person zero-sum game there are strategies such that the maximum of the minimum gain for one player is equal to the minimum of the maximum loss for the other. In some cases these strategies are pure, but usually they are mixed. In other words, an equilibrium exists for games of this kind. The implication, that competitive situations can lead to a fair and balanced outcome, inspired von Neumann and Oskar Morgenstern to write their famous book, *The Theory of Games and Economic Behavior*, first published in 1944. Since then a great deal of work has been done on extending the idea of equilibrium and applying it to other kinds of games, such as games in which the players do not have complete information about the outcomes.

The law of large numbers

It is believed that the origins of coined money, and the deductive approach to mathematics, can both be ascribed to the Greek world, in particular the part of Asia Minor known as Lydia. Writing in about 450 BC, Herodotus claimed that the Lydians had also invented many games of chance, to distract themselves while they were enduring a great famine. That is almost certainly false but, as we have seen, gambling has long been a common pastime. Indeed there is ample evidence that a primitive form of gambling, based on the outcome of a throw of a small bone, was popular in the Roman era. One such bone was an astragalus, which is found in the heels of sheep and other animals. Its irregular shape means that when an astragalus is thrown it can land in one of four positions, but they are not all equally likely. Although the outcome of a single throw cannot be predicted, there is nevertheless a certain consistency when a large number of throws is made, because the proportion of throws resulting in a particular outcome seems to remain roughly the same, however often the experiment is repeated. It is not clear from the Roman evidence whether this observation was ever formulated clearly in numerical terms. However, it is reasonable to suppose that professional gamblers must have realized that profit can be made by knowing the characteristics of a particular astragalus.

In the ninth century AD a similar principle was used by Arab cryptographers, to decrypt messages that had been encrypted by the method of substitution (Chapter 5). Suppose a certain symbol, call it X, occurs m times in a piece of text with

n symbols. It is an observed fact that the relative frequency m/n remains fairly constant, whatever the text, and so its value can be assumed known (approximately), for any text. The relative frequency m/n can thus be interpreted as an estimate of probability. If we look at any specific place in the text, say the 57th symbol, then the probability of the event that this symbol is X is m/n.

When the notion of probability was formally defined in the middle of the seventeenth century, the resulting mathematical framework helped to clarify the relationship between mathematical models and practical applications, such as gambling and cryptography. Jakob Bernoulli in his *Ars Conjectandi* (1713), and Abraham de Moivre in his *Doctrine of Chances* (1718, second edition 1738), were the first to consider probabilistic models of actual situations, such as tossing a fair coin. They imagined a process that produces, at each trial, one of two outcomes, H (heads) and T (tails), each with probability 1/2. The outcome of a sequence of trials produced by this process will look something like

HHTHTTHTHTTTTHHTTHTHTHHTTTHTHTHHHTTHTHTHTTHHH.

If there are 100 trials then there are 2^{100} possible outcomes, and each one has probability $(1/2)^{100}$. Intuitively, we expect that the sequence will contain about 50 heads and about 50 tails, although we should not be surprised to find that the numbers are not quite equal. Bernoulli and de Moivre set out to provide mathematical justification for this intuition. What is the probability of exactly 50 heads and 50 tails—in other words, how many of the 2^{100} possible outcomes have this property? More generally, what is the probability of exactly h heads and $100 - h$ tails?

The total number of possible sequences, 2^{100}, is a number with over 30 digits, so this is not a problem that we can solve by listing all the possibilities. However, we can make a little progress by using some elementary combinatorics. The number of sequences with h heads in n trials is the number of ways of choosing h 'successful' trials from the total number n, and this is just the binomial coefficient $\binom{n}{h}$.

Dividing this by the number of possibilities, 2^n, gives the required probability. For example, the probability of exactly 50 heads in 100 trials is

$$\frac{1}{2^{100}}\binom{100}{50}.$$

It is important to understand why this formula is not the complete answer to our problem. A formula is simply a compact way of representing an algorithm, a method for computing a number, and its usefulness depends on how easy it is to carry out those computations. This formula determines a definite numerical value, and it tells us how to work it out in theory, but in practice it is not much help. According to Bhāskara's formula for the binomial coefficient, we know that

$$\frac{1}{2^{100}}\binom{100}{50} = \frac{1}{2 \times 2 \times \ldots \times 2} \times \frac{100 \times 99 \times 98 \times \ldots \times 51}{1 \times 2 \times 3 \times \ldots \times 50}.$$

Evaluating this expression involves lots of multiplications, and if you try it, you will soon find out why there is a problem.

De Moivre's method was based on two of the tools that had been developed in the seventeenth century, logarithms and infinite series. The first step was to apply the basic property of the logarithm, the fact that the logarithm of a product like $1 \times 2 \times \ldots \times n$ is the sum of the logarithms of the terms, $\log 1 + \log 2 + \ldots + \log n$. (Note that $\log 1$ is actually 0.) Then, in the manner of Newton, he expressed each logarithm as an infinite series, and after some bold algebra, he found that

$$\log 1 + \log 2 + \ldots + \log n \quad \text{is approximately} \quad \left(n + \frac{1}{2}\right)\log n - n + c,$$

where c is a number that he calculated to be about 0.919. Armed with this formula, and variations of it, de Moivre was able to obtain estimates for the probabilities that arise in the coin-tossing problem. One of his first results was that the probability of exactly 50 heads in 100 trials is about 0.079559.

At first sight, de Moivre's results may appear to be no more than a neat way of doing some nasty calculations, but they foreshadowed a new way of relating mathematics to real-life situations. Suppose, for simplicity, that n is an even number. Then it seems reasonable that the most likely outcome of n trials is that the number h of heads is $n/2$ and the number of tails is also $n/2$. Similarly, it seems reasonable that the probability of $n/2 + 1$ heads and $n/2 - 1$ tails will be only slightly less, and that generally the probability of $n/2 + d$ heads and $n/2 - d$ tails decreases as d increases.

De Moivre was able to prove these assertions mathematically. Furthermore, his numerical estimates allowed him to calculate the probability that h lies within a given range. For example, he found that 'if 3600 Experiments be taken . . . then the Probability of the Event's neither appearing oftener than 1830 times, nor more rarely than 1770, will be 0.682688'. Statements like this one lead to a result that has become known as the Law of Large Numbers. De Moivre's results imply that the probability that h/n differs from 1/2 by more than any given amount approaches zero as n increases. This statement can be regarded as a mathematical justification for our intuitive notion of probability: in this case, the probability of a head in any one trial can be measured experimentally by the relative frequency h/n of h heads in n trials. But it must be stressed that the law of large numbers is open to misinterpretation: it is not a universal justification for glib statements that certain things are bound to happen 'on average'. And, as we shall see, its application to real-life situations is usually dependent on hidden assumptions, especially when those situations involve human beings.

We are on firmer ground when we look at the mathematical consequences of de Moivre's work. One amazing discovery was made by the Scottish math- ematician James Stirling, who had independently worked on sums of logarithms like $\log 1 + \log 2 + \ldots + \log n$. He found that the number c in de Moivre's approximation $(n + 1/2) \log n - n + c$ is the logarithm of the square root of 2π.

The appearance of π, the area of a circle with unit radius, in the context of calculations about probability and binomial coefficients, must have been a great surprise. Even more significant is that the number e also plays a part. With hindsight, this is less surprising, because we are now familiar with the fact that e^x is another name for 'the number whose hyperbolic logarithm is x'. However, this notation was not used by de Moivre and Stirling in the 1730s. When de Moivre's approximation is expressed in terms of e it becomes a formula for what we now call the *factorial* of n, the product $1 \times 2 \times \ldots \times n$. It is usually written as $n!$ and the result is known as Stirling's Formula:

$$n! \quad \text{is approximately} \quad \sqrt{2\pi n}\left(\frac{n}{e}\right)^n.$$

This is one of the most useful tools in the mathematical theory of probability.

Statistics

The collection of data for administrative purposes is an ancient practice. Censuses were regular events in Imperial Rome, and they were used mainly to ensure that the citizens rendered unto Caesar what he claimed to be his, by way of taxation and military service. A thousand years later the Norman conquerors of England compiled the *Domesday Book* for similar reasons. The officials who used this information were undoubtedly able to draw some general conclusions from it, but there is no evidence that they organized and consolidated the data in any way, for example by calculating averages. Progress in this respect was slow, and it was not until the sixteenth century that we can find traces of what we should now call statistical methods.[7] Tables of 'vital statistics' were an important tool in the business of annuities and life insurance, and their calculation inevitably involved making deductions from a mass of empirical data.

At the beginning of the eighteenth century the work of Jakob Bernoulli and Abraham de Moivre led to a better understanding of the implications of observed data. For example, when de Moivre's numerical estimates for the probability of the number h of heads in 100 tosses of a fair coin are plotted, we obtain a distinctive picture (Figure 8.3). The probability is greatest when $h = 50$ and falls away on either side of this value, becoming very small (but not quite zero) at the extreme values $h = 0$ and $h = 100$.

A similar picture is obtained for any number of trials n, and as the number increases the plotted points approach a smooth curve, known as a *bell curve*. The equation of this curve can be expressed neatly in terms of e^{-d^2}, where $d = h - n/2$ is the deviation from the maximum.

The fundamental importance of the bell curve was not recognized until the early years of the nineteenth century. At that time there was great interest in the use of astronomical observations to predict the movements of planets, comets, and (in particular) asteroids. But empirical observations were subject to many errors, and

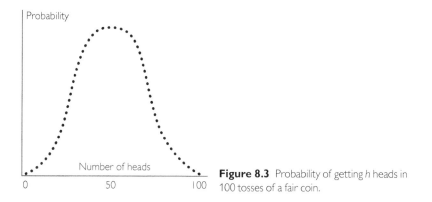

Probability

Number of heads

0   50   100

Figure 8.3 Probability of getting *h* heads in 100 tosses of a fair coin.

it was necessary to take several readings and combine them to get a reliable one. What was the best way of doing so? Legendre (1805) and Gauss (1809) discovered a neat solution to this problem, and Gauss justified it by assuming that the probabilities of observational errors could be described by a bell curve. This assumption received mathematical support in 1812, when Pierre-Simon Laplace published his *Théorie Analytique des Probabilités*. The bell curve, in its analytic form, played a leading part in his work. Laplace suggested that a numerical observation can be considered as the average of a large number of independent random contributions (errors). He proved that the distribution of such a variable will tend towards the bell curve, whatever the distribution of the errors. This result is now known as the Central Limit Theorem, and together with the Law of Large Numbers, it underpins much of statistical theory.

As the study of statistics progressed, it became clear that the bell curve is a good model for many other kinds of data. For this reason the corresponding distribution of probabilities became known as the *normal* distribution. In some cases the use of the normal distribution is justified because Laplace's Central Limit Theorem seems to apply, but in others its use is essentially based on conjecture. This can lead to problems; for example, the economic crises of the early twenty-first century have been blamed on the assumption that some financial variables are normally distributed (see Chapter 10).

More generally, in statistical matters the link between theory and practice is often fraught with difficulty. To draw a conclusion from observed data, it is important to be clear what assumptions are being made. A good illustration arose in our discussion of the Problem of Points. Recall that Pacioli had suggested that the stakes should be divided according to the score when play is interrupted, which corresponds to the hypothesis that the current score provides accurate information about the relative skill of the players. However, this is open to doubt. After one game the score must be 1–0 or 0–1, implying that one player is infinitely better than the other. That is not usually the case, because the very fact that they have agreed to play against each other, and gamble on the result, suggests that they have comparable skills. Fermat's rule for dividing the stakes was based on the hypothesis that the players have equal skills, which is usually a better assumption in practice.

Another example occurs in the coin-tossing problem. We considered the 'fair coin' as a mathematical ideal, and assigned the probability 1/2 to the two outcomes H and T. When we apply the theory to a real coin, we are making the hypothesis that it is fair, in the sense that the outcomes H and T are equally likely. Of course a real coin may not be fair; if it is slightly damaged then its physical properties do not correspond exactly to the ideal. If the experiment with 100 tosses is repeated many times, and each time we obtain about 70 heads, then we have good reason to suppose that our coin is not fair.

These examples only skim the surface of the deep ocean that is the theory of statistical inference. In the twentieth century statisticians developed good techniques to allow them to deal with practical situations, and these rules have been used for the great benefit of humanity. The advances in medicine that have saved millions of lives, and prolonged billions more, could not have been made without statistical analysis. But there is still a serious debate on exactly how the mathematical notion of probability should be interpreted. One group of statisticians, known as the Bayesians, hold that the probability of an event is based on a 'degree of belief', rather than the notion of relative frequency. They take their name from an elementary result of Thomas Bayes, published posthumously in 1763. In the words of Bayes: 'the probability that two ... events will both happen is ... compounded of the probability of the 1st and the probability of the 2nd on supposition that the 1st happens'. Laplace, in his 1812 *Théorie*, was the first to recognize the importance of this principle in statistical inference.

In abbreviated form, Bayes' result says that, for any two events A and B,

$$Probability(A \text{ and } B) = Probability(A) \times Probability(B \text{ given } A).$$

The last term is now known as the *conditional* probability of the event B, given that the event A occurs. For example, suppose A is the event that a dice shows an even number, which has probability 1/2, and B is the event that it shows 4, which has probability 1/6. Then the result tells us that the probability of a 4, given that the number is even, is 1/3. No one disputes the truth of Bayes' result; the debate is about its status in the theory of statistical inference.

One other eighteenth-century innovation in the field of statistics is worthy of note. Statistical data, ranging from numbers of births to trade figures, was traditionally presented in undigested form, as tables of numbers. In 1786 William Playfair conceived the idea of presenting observed data in pictorial form, in the same way as curves were plotted in Cartesian geometry. In his view 'making an appeal to the eye when proportion and magnitude are concerned is the best and readiest method'.

Working on this principle, Playfair constructed diagrams of English overseas trade by plotting the monetary value of imports and exports against the time in years (Figure 8.4).[8]

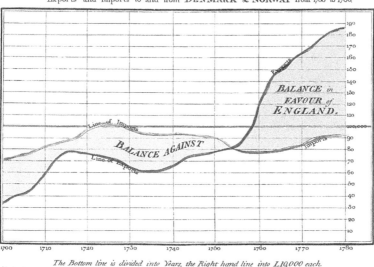

Exports and Imports to and from DENMARK & NORWAY from 1700 to 1780

BALANCE in
FAVOUR of
ENGLAND.

Line of Imports

BALANCE AGAINST

Line of Exports

Imports

The Bottom line is divided into Years, the Right hand line into £10,000 each.

Figure 8.4 One of Playfair's charts of the balance of trade. William Playfair, *Commercial and Political Atlas* (London, 1786). Diagram from Wikimedia Commons.

There is something of a mystery as to why these pictures of financial data were not adopted more widely at the time. It may have been the cost of preparing the diagrams for printing, or possibly it was a reaction against Playfair's dubious reputation as an entrepreneur. In Chapter 10 we shall see that the pictures did catch on eventually, and led to significant advances.

CHAPTER 9

Modelling and Measuring

You are often inclined to 'do your own thing'. In an enlightened society there are certain norms and conventions, designed to accommodate your individual preferences within a framework that works for the common good. Economists and political scientists have tried to find ways of describing these matters in mathematical language. In this chapter you will find that progress was closely linked, in time and place, with an apparently quite different enterprise—the construction of common standards of weight and measure. As we saw in Chapter 2, that problem goes back to the first stirrings of mathematical activity.

Mathematics in context

The idea of giving names to numbers, and using the names to count objects like stone tools and sheep, first arose in the period for which we have no written records. Methods of working with numbers are inextricably linked with the next stage of development, when human beings began to live and work in a community and written records were essential. So it is clear that, almost from the very beginning, mathematics has been part of the social context. For example, the need to allocate resources equitably led to the development of arithmetical procedures for dividing one number into another. There were different social systems in different parts of the world, but calculation played an essential part in all of them.

In the Greek, Roman, and Islamic periods, the increasing complexity of the social structure led to the development of better methods of calculation, particularly in matters relating to coins, weights, and measures. But a new kind of mathematics also began to take root. The idea that mathematical procedures are Right—that they embody a special kind of truth—had been taken for granted in the case of arithmetic, and the Greeks discovered that geometry required a more careful approach. They also began to study properties of numbers, using logical arguments rather than intuitive observations. Thus there arose the notion that mathematics could be studied for its own sake. Nevertheless, for nearly 2000 years after Euclid, most advances in mathematics were motivated by practical problems, such as navigation. It was not until the early modern period that 'pure' mathematics really began to grow, nourished by the symbolic methods of algebra. Great mathematicians like Newton and Gauss were often motivated to use the new methods of calculus to solve problems that arose in the 'applied' side of the subject. The development of a mathematical theory of probability added a new twist to the relationship between theory and practice. Debate about the theoretical foundations of probability and statistical inference continues to this day, even though we rely on the subject for many important decisions. Indeed, the entire subject of mathematics 'for its own sake' has turned out to be a great benefit to humanity, because arithmetic, geometry, algebra, calculus, and probability have been used to solve a wide range of practical problems, sometimes quite unexpectedly. In this chapter we shall explain how common problems of economics and politics came to be studied by mathematical methods, and how the development of the international community of science led to advances in the art of measurement.

Measuring value and utility

Money takes many forms and has many functions. One of its purposes is to serve as a common measure of value, so that the associated money-objects, such as coins and cowrie shells, can be used to facilitate the exchange of goods. However, a common measure of value cannot be taken for granted. At the start of this book the scenario of the slaughtered mammoth was used to illustrate that, in prehistoric times, the value of a commodity could depend on a person's need for it. In that scenario the cut-and-choose algorithm provided a means of resolving a dispute without the use of money. When the growth of trade and industry led to the widespread use of money, it was inevitable that questions would arise about the relationship between the value of an object and its monetary counterpart, the price.

The ancient philosophers speculated about these matters in general terms, but it was not until the eighteenth century that people began to think numerically about the measurement of value. The new theory of probability seemed to offer some insights, and the fourth part of Jakob Bernoulli's *Ars Conjectandi* contained an attempt to apply this theory to civil, moral, and economic affairs. But the book was not published until 1713, eight years after Jakob's death, and it appears that it contains only a sketch of what he was trying to achieve. Around 1730 Daniel, the son of Johann Bernoulli, took up the subject again. In 1738 his paper on 'a new theory of risk evaluation' was published. It began with a scenario that was rather more modern than the slaughtered mammoth.

> A poor man obtains a lottery ticket that will yield with equal probability either nothing or twenty thousand ducats. Will he evaluate his chances of winning at ten thousand ducats? Would he not be ill-advised to sell his ticket for nine thousand ducats? To me it seems that the answer is negative.

On the other hand, said Daniel, a rich man would be well advised to buy the ticket, because his needs are different.

Bernoulli continued with a more general discussion of the concept that we now call *utility*. He considered how the utility of a commodity, such as apples, can depend on the amount available. The utility U of a number x of apples clearly increases with x, but the *rate* of increase will *decrease* as x becomes larger, because we can only eat so many apples. In particular, he suggested that the formula $U = \log x$ might be used to represent this situation, because it has the right properties. As shown in Figure 9.1, the values increase, but the rate of increase gradually decreases.

Bernoulli assumed that the values of x are whole numbers, or, as we now say, x is a *discrete* variable, rather than a *continuous* one. In that case the rate of increase of U at a specific value of x is measured by the difference between the values of U at x and $x - 1$. Economists now refer to this as the *marginal utility* of U and denote it by MU; they sometimes describe it as 'the utility of the last apple'. Here the change in U, which we may denote by dU, is being measured with respect to the change of one unit in x; in other words, $dx = 1$. In the case of a continuous variable x, MU corresponds to dU/dx. This means that the differential calculus can be deployed. Indeed, Daniel Bernoulli used this analogy to justify his choice of a logarithmic curve to represent utility.[1]

In the latter part of the eighteenth century the classical school of economists and philosophers turned their attention to questions of utility, value, and price. Adam Smith attempted to elucidate the meaning of value, but his analysis was based on notions that he did not explain convincingly. Jeremy Bentham discussed utility in far more general terms, with the aim of making interpersonal comparisons that could provide a rational basis for legislation. But the possibility of using mathematics to explain relationships between economic variables was not explored effectively for a hundred years after Bernoulli's first attempt.

Antoine Augustin Cournot was appointed as Professor of Mathematics at the University of Lyon in 1834. He was familiar with the tremendous advances that had been made in the theory and applications of the differential calculus and, in his *Researches into the Mathematical Principles of the Theory of Wealth* (1838), he

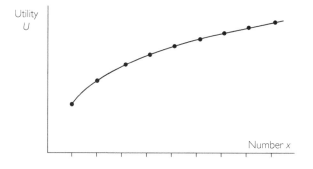

Figure 9.1 A utility curve of logarithmic form.

had no qualms about representing economic variables by algebraic symbols. Thus, he used p for the price of a commodity and D for the demand for it. He expressed the economic fact that there is a relationship between D and p in mathematical language, by saying that D is a *function* of p and writing $D = F(p)$. By using F, an unspecific general function, rather than a formula, he laid the foundation for a theory of economic activity based on the mathematical analysis of functional relationships. For example, the economic fact that demand falls as the price increases can be translated into mathematics by saying that the rate of change dF/dp is negative. The differential calculus is well-suited to such analysis, because many economic activities can be regarded as problems in optimization. A simple example is the problem of a firm that wishes to set the price of its product so that its revenue is maximized. If the price is p and the demand is D, the revenue is $p \times D = p \times F(p)$. The differential calculus tells us that the maximum of $p \times F(p)$ occurs when its rate of change is zero, and this fact determines the optimum price.[2]

A few years after the publication of Cournot's paper another Frenchman, Jules Dupuit, wrote an article entitled 'On the measurement of utility in public works'. Among other things, he described how the number of people using a toll-bridge might depend on the toll charged, and he explained how data (albeit hypothetical) could be used to measure the benefit derived by the public. Dupuit's data was discrete, but in an Appendix he drew the curve of demand in terms of a continuous variable (Figure 9.2). He clearly pointed out that his analysis could be expressed in the language of the calculus:

> for convenience of exposition we have calculated differences instead of using the differential calculus. Those who are familiar with the elements of the calculus will see later how precision may be substituted for approximation.

A few decades after the publication of Dupuit's work several economists began to use the differential calculus to investigate the relationship between utility and price. Their work was based on the scenario of a consumer (Jenny) who has a fixed budget to spend on several goods, given the prices of the goods and her utilities for

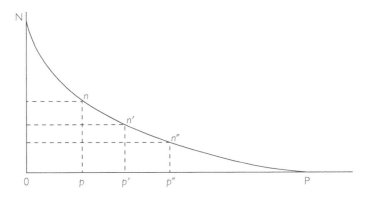

Figure 9.2 Dupuit's curve showing demand as a function of price.

them. This can be formulated as an optimization problem: Jenny wishes to maximize her utility, subject to the constraint that her budget is fixed. It is easy to make a mathematical model of this situation. Suppose there are just two commodities, with prices p_1 and p_2 dollars per unit, and the budget is m dollars. The purchase of x_1 units of the first commodity and x_2 units of the second commodity costs $p_1x_1 + p_2x_2$ dollars, and so x_1 and x_2 must satisfy the constraint $p_1x_1 + p_2x_2 = m$. In geometrical terms, this means that the point (x_1, x_2) must lie on a straight line, the *budget line*, as depicted in Figure 9.3(1).

Now suppose that Jenny's utilities for the quantities x_1 and x_2 are $U_1(x_1)$ and $U_2(x_2)$, respectively. Then the problem is:

$$\text{maximize the total utility } U_1(x_1) + U_2(x_2)$$
$$\text{subject to the budget constraint } p_1x_1 + p_2x_2 = m.$$

Among the economists who worked on this problem in the 1870s were Stanley Jevons, Carl Menger, and Léon Walras. By a combination of mathematical and economic arguments, all three obtained similar solutions. Soon afterwards, Francis Ysidro Edgeworth gave an exposition in his book, *Mathematical Psychics*, published in 1881.

Edgeworth's approach was geometrical, and was based on the curves defined by the rule that $U_1(x_1) + U_2(x_2)$ is constant. At all points on a fixed curve of this kind the total utility is the same, and we can say that Jenny is 'indifferent' as to which point (x_1, x_2) on the curve is chosen. So Edgeworth called it an *indifference curve*. Sketching the indifference curves for a few typical values of the constant we get the general picture, as in Figure 9.3(2).

Jenny's problem is to find the largest value of the constant for which the indifference curve has a point of intersection with the budget line. For larger values of the constant, the indifference curve will not intersect the budget line: Jenny cannot afford that level of utility. On the other hand, for smaller values of the constant, there will generally be two points of intersection. The diagram suggests that the best value is the one for which there is just one point of intersection; specifically, the point where the budget line is tangent to an indifference curve (Figure 9.3(3)).

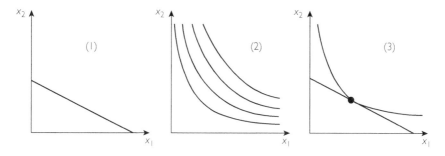

Figure 9.3 The consumer's problem: (1) the budget line; (2) indifference curves; (3) the optimum.

Using elementary calculus Edgeworth was able to explain the result obtained by Jevons, Menger, and Walras. His argument went as follows. The slope of the budget line is $-p_1/p_2$, and this line is tangent to the required indifference curve. So we need a formula for the slope of an indifference curve. The equation of the curve defines x_2 implicitly in terms of x_1, and by the rules of the calculus the rate of change of x_2 with respect to x_1 is

$$\frac{dx_2}{dx_1} = -\frac{dU_1}{dx_1} \bigg/ \frac{dU_2}{dx_2}.$$

We need the particular indifference curve for which the point with this slope lies on the budget line. Hence the optimum point can be found by solving the two equations

$$\frac{dU_1}{dx_1} \bigg/ \frac{dU_2}{dx_2} = p_1/p_2 \quad \text{and} \quad p_1 x_1 + p_2 x_2 = m.$$

Rearranging the first equation, and translating the derivatives into the economists' marginal notation, it follows that the optimum allocation (x_1, x_2) is such that

$$\frac{MU_1(x_1)}{p_1} = \frac{MU_2(x_2)}{p_2}$$

In other words, the optimum quantities are such that the ratio of the marginal utility to the price is the same for each good. For example, if Jenny's budget is one dollar and she decides to buy 4 apples at 15 cents each and 2 bananas at 20 cents each, then her desire for the fourth apple and her desire for the second banana are in the ratio 15:20. Possibly the first person to enunciate this principle was Heinrich Gossen, a writer of some deep, but very obscure, musings on economic theory. In 1854 he wrote that

> a person maximizes his utility when he distributes his income among the various goods so that he obtains the same amount of satisfaction from the *Geldatom* spent upon each commodity.[3]

The introduction of calculus led to a sea-change in economic thought, often called 'the marginal revolution', and in due course the subject that goes by the name of neoclassical economics emerged. The term is not strictly defined, but one of its characteristic features is the attempt to set up models of economic behaviour that can be based on mathematical relationships. For example, the discussion of an individual consumer outlined above can be extended to throw light on situations involving several competing consumers. Edgeworth himself used his indifference curves to analyze the case of two persons marooned on a desert island who wish to share out their provisions equitably. He found that there are many solutions to their problem, and they can be represented pictorially as the points on what he called a *contract curve*. Much later, when the Theory of Games was developed mathematically, people realized that the same idea could also be expressed in that framework.

The problems of democracy

It is a truism that different individuals value goods in different ways. John may consider that one apple is worth two bananas, but Paul may think that an apple and a banana have equal value. As we know, difficulties can arise when John and Paul are required to combine their preferences. Although we try to rely on a democratic vote to resolve such situations, there are many reasons why that is not a complete answer. The traditional procedure is for voters to vote only for their most preferred option, and the option that receives most votes wins. But this procedure may not always select the best option, in the opinion of the electorate as a whole.

Towards the end of the eighteenth century the problem of voting was discussed by several French mathematicians. The first of them was Jean-Charles de Borda, a man of many parts. His work as a military engineer had led to his appointment to the French Academy of Sciences in 1770. He also served as a captain in the French navy, and was captured by the English, but he returned to France and played a prominent part in the affairs of the Academy. In 1770 Borda had written about the situation where voters have to elect one candidate to a fill a position; his paper was printed in a volume of the *Histoire de l'Académie* dated 1781 (but not published until 1784). His analysis was based on the observation that each voter has a ranking order for the candidates, but casting a single vote only partly represents that order. Borda considered the example of three candidates A,B,C, where there are six possible ranking orders, corresponding to the six permutations ABC, ACB, BAC, BCA, CAB, CBA. Suppose there are 21 voters, and their preferences are distributed as follows:

First preference	A	A	B	B	C	C
Second preference	B	C	A	C	A	B
Third preference	C	B	C	A	B	A
Number of voters	3	5	0	7	0	6

Using the standard procedure, each voter will cast a vote for their most preferred candidate, so A gets 8 votes, B gets 7, and C gets 6, which means that A wins. But, as Borda pointed out, most voters (13 out of 21), actually prefer either B or C to A.

Borda suggested that the difficulty could be overcome by assigning marks to each candidate, based on their position in the ranking. If we assign 3 marks for a first preference, 2 marks for a second preference, and 1 mark for a third preference, the total marks for A, B, and C are 37, 44, and 45, respectively. On that basis, C will be elected. This procedure is known as the *Borda count*. It was used to elect members of the Academy until 1801, when it was abandoned on the insistence of that famous democrat, Napoleon Bonaparte. It is still in common use today, in a variety of circumstances.

The Borda count is not beyond criticism. For example, it assumes that the steps in the relative preferences of voters are uniform; so that when the ranking is ABC, A is preferred to B by exactly the same amount as B is preferred to C. It is fairly easy to devise procedures for overcoming such difficulties, but unfortunately it seems that it is always possible to invent a scenario where the procedure leads to a perverse outcome.

In 1785 Borda's friend the Marquis de Condorcet tried to apply the theory of probability to electoral procedures in a long and complicated *Essai*. Most writers on probability found the *Essai* hard to understand, and sometimes inconsistent, but Condorcet had indeed uncovered many of the complexities of the subject.[4] Slightly later, Laplace also tried to apply probability theory, and he produced a theoretical proof of the superiority of the Borda count, under certain assumptions. But he too realized that the vagaries of human behaviour, such as tactical voting, might invalidate his conclusion.

A century later, the theory of elections was discussed by Charles Dodgson, more famous as Lewis Carroll, the author of *Alice's Adventures in Wonderland*. In real life he was a lecturer in mathematics at Christ Church, Oxford, and Alice was one of the daughters of the head of the college, Dean Liddell. His interest in elections was aroused by various contentious issues that came before the Governing Body of the College, and was fanned by animosity towards Liddell. In the years 1873−8 he wrote several pamphlets on electoral procedures, and covered some of the ground trodden by Borda, Condorcet, and Laplace, but he made no significant advance.[5]

In the twentieth century the rise of mathematical economics led to much interest in the relationships between preferences and utilities. The so-called *cardinal* utility theory, the notion that a person's need for a bundle of goods can be expressed by assigning a number, was criticized as being artificial. Attempts were made to replace it by an *ordinal* approach, based on ranking by means of preferences. In their *Theory of Games and Economic Behavior* (1944), von Neumann and Morgenstern tried to reconcile the two approaches. They formulated some conditions which ensure that a preference ordering can be represented by a utility function U which assigns a number $U(X)$ to the action X in such a way that whenever X is preferred to Y then $U(X)$ is greater than $U(Y)$. But to do this they were forced to introduce some axioms that are somewhat artificial.

The suspicion that there are fundamental obstacles to basing 'democracy' on the aggregation of individual preferences was confirmed in the early 1950s. Kenneth Arrow formulated some plainly reasonable conditions that should hold when individual preferences are combined into a single *social preference*, and he proved that there is no way that they can be satisfied. His conditions were later refined slightly, and can be stated quite simply as follows.

1 If every voter ranks A above B then their social preference ranks A above B.
2 If a set of voters' preferences is such that their social preference ranks A above B, and some voters revise their preferences, but without changing their views on the relative merits of A and B, then the new social preference also ranks A above B.
3 There is no dictator—that is, no one person's preference will always correspond to the social preference.

The fact that there can be no general method of defining a social preference that satisfies these conditions is known as *Arrow's Impossibility Theorem*.

The measure of the world

The interest in democratic procedures shown by the French mathematicians of the late eighteenth century was a reflection of their nation's turmoil at that time. Borda, Condorcet, and Laplace all became involved in the events of the French Revolution, and in doing so they laid the foundations for a different kind of revolution—in measurement.

By this time the need for standards of weight and length was acknowledged in all the developed nations. In 1742–3 the Academy in Paris and the Royal Society in London exchanged copies of their national standards, and compared them with extreme care. They were particularly interested in the so-called troy weights, used for gold and silver items, because these were also used in scientific experiments. They found that the ratio between the troy-weight standards of France and England was exactly 64:63. However, in both countries the position in the wider community was not so straightforward. In the British Isles the bushel measure of capacity varied from place to place, while in France the lingering medieval system of government meant that all kinds of weights and measures were determined locally. The French systems were particularly resistant to change, because doing so would affect the feudal rights of the aristocracy. Some revolutionary leaders took up this issue, and in April 1790 the matter was raised in the General Assembly by Charles-Maurice de Talleyrand. He proposed setting up a system of weights and measures that would not only be uniform throughout France, but would also be truly international, because it was based on objective standards. A committee was appointed to consider the matter, three of the committee being the aforementioned Borda, Condorcet, and Laplace. In March 1791 they reported on how the standard unit of length might be defined objectively, and they concluded that it should be based on the circumference of the earth. Specifically, they recommended that it should be one ten-millionth part of the length of a meridian from the North Pole to the equator. The proposed unit should be named the *metre*.

The work began in 1792. The plan was to determine, by the traditional method of triangulation, the length of a section of the meridian that runs through Dunkirk and Barcelona (Figure 9.4). When the length of the baseline is known, and the angles of the triangles have been measured, the distance between the end points can be calculated. The latitudes of the end points can then be found by astronomical methods, giving the required answer. The theory was relatively easy, but there were enormous practical difficulties.[6] The surveyors had to operate in difficult terrain, and they were often harassed by suspicious local people, who thought they might be spies. This was not entirely unreasonable, given that a revolution was in progress. The lengths and angles had to be measured with extreme accuracy.

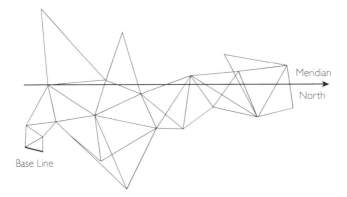

Figure 9.4 The triangulation used for surveying the meridian near Carcassonne.

Borda himself constructed measuring rods, and an instrument known as a 'repeating circle', which facilitated the measurement of angles.

Borda was also involved with constructing the new standard of weight. It was agreed that the unit, to be known as the *gramme* (now usually written *gram*), should be the weight of a volume of pure water equal in size to a cube with sides of one-hundredth of a metre. As this was impractical as a physical definition, it would be represented by a standard object, a kilogram (equal to 1000 grams) made of a suitable metal.

All this work was expected to take some years, as indeed it did. Fortunately there were many people who thought it worthwhile, even in a time of national turmoil. The Reign of Terror in 1793 led to the closing of the Academy, but the commission on weights and measures was allowed to continue its work. Condorcet, who had been a fervent supporter of the revolution, was imprisoned and died, and Laplace thought it wise to leave Paris for a while. He was thus able to continue writing his great work, the *Mécanique Céleste*, in which Newton's geometrical arguments were expressed in the language of analysis and algebra. Despite such setbacks, the establishment of the new Metric System was authorized by a decree issued in 1795 (Figure 9.5).

Sadly, the adoption of the Metric System did not begin well. Partly this was because some of the innovations were so awkward that failure was almost inevitable. The convenience of decimal division for the units of length and weight was clear, but the proposed application of decimals to the calendar was thoroughly confusing. The year was supposed to be divided into 12 months, each of 3 weeks, each week was 10 days, each day 10 hours, and each hour 100 minutes, and each hour 100 seconds. A slightly more reasonable proposal was to measure angles in terms of a new degree, 100 of them making a right-angle, instead of the old degree, 90 of which made a right-angle. But this too was a failure. The most serious problem was the innate opposition of most of the French people, whatever their revolutionary zeal, towards any new form of measurement.

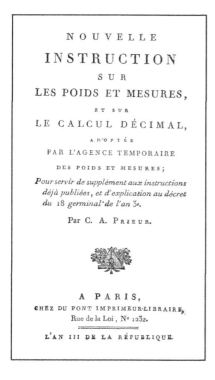

NOUVELLE

INSTRUCTION

S U R

LES POIDS ET MESURES,

ET SUR

LE CALCUL DÉCIMAL,

ADOPTÉE

PAR L'AGENCE TEMPORAIRE

DES POIDS ET MESURES;

*Pour servir de supplément aux instructions
déjà publiées, et d'explication au décret
du 18 germinal de l'an 3e.*

Par C. A. PRIEUR.

A PARIS,

CHEZ DU PONT IMPRIMEUR-LIBRAIRE,
Rue de la Loi, N° 1232.

L'AN III DE LA RÉPUBLIQUE.

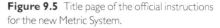

Figure 9.5 Title page of the official instructions
for the new Metric System.

For that reason it was decreed in 1812 that the Metric System could be used in conjunction with some of the older units of length and weight. The result was a complete muddle, perhaps a true reflection of the governance of France at the time. The restoration of the Metric System in 1840 was a long overdue return to sanity, and after that there would be no more bureaucratic bumbling. In due course the Metric System was adopted by most nations of the world, thus vindicating the vision of the French mathematicians and scientists who set it up. A few countries retained their traditional quaint terminology for units of measurement, but most of them (including the UK since 1963) have now chosen to define these units in terms of the international metric standards.

It is somewhat ironic that the success of the Metric System did not, ultimately, depend on the tremendous effort that went into determining the length of the meridian. It gradually became clear that this length cannot be used as an objective standard, because the Earth is not a perfect sphere. Metallic standards of the metre and the kilogram had been constructed in 1799, and for many years it was thought that the object known as the 'metre of the archives' could serve adequately as a primary standard of length, irrespective of its relationship with the length of the meridian. However, in due course that plan had to be abandoned too, because the object itself showed clear signs of deterioration. Consequently, new standards of the metre and the kilogram were made, using a highly stable alloy of platinum and iridium, and these were formally adopted in 1901.

The system of having unique primary standards, from which all others were derived, was fine in theory, but it created some problems in practice. The very act of comparing a secondary standard with the primary one was likely to affect both of them, and even a tiny change was not acceptable. So in the twentieth century the scientific community began to reconsider the possibility of objective standards. The principle was that they should be based on physical phenomena that would give precisely the same result wherever they were observed. In 1960 the metre was defined in terms of the wavelength of light emitted by the isotope krypton-86, and in 1967 the unit of time (the second) was defined in terms of a clock based on the isotope caesium-133. The latter definition paved the way for a further re-definition of the metre in 1983, in terms of the speed of light, one of the most basic physical constants. So now the metre is officially the distance travelled by light in a vacuum in one 299792458th part of a second.[7]

Unfortunately, attempts to define the kilogram by a reproducible scientific method, based on fundamental physical constants, have met with only partial success.[8] Consequently, in 2015 the International Prototype Kilogram from 1901 is still the primary standard of mass and weight, and is surely one of the most important objects on our planet.

CHAPTER 10

Mathematics and Money in the Age of Information

W hen you pay your bills using a plastic card, you are simply authorizing alterations to the information stored in some computers. This is one aspect of the symbiotic relationship that now exists between money and information. The modern financial world is byzantine in its complexity, and mathematics is involved in many ways, not all of them transparently clear. Fortunately there are some bright spots, such as the fact that it is now possible to measure information in a mathematically precise way.

Financial speculation

The idea of speculating on the basis of a belief in future events can be traced back to the primitive industries of prehistoric times. Before 3000 BC extensive underground tunnels were constructed in the east of England, at a place we call Grime's Graves. Despite their name, the tunnels were not intended for burying the dead, but for mining flint, from which the stone tools mentioned in Chapter 1 of this book were made. We do not know how the production and distribution of the flint was organized, but it is clear that some degree of speculation must have been involved. The people who did the mining, in extremely arduous conditions, invested their labour in the expectation that the material they produced would somehow provide them with the necessities of life.

By the seventeenth century trade and commerce formed an important part of the monetary economy of England, and wealthy people were able to speculate by investing money, rather than labour. Groups of 'merchant adventurers' like the East India Company allowed investors to buy shares, which offered a return in the form of a dividend, if the company prospered. The shares themselves could also be traded, so that people could speculate in the hope that their shares might eventually be worth more than they paid for them. Isaac Newton's position as Master of the Mint had made him a wealthy man, and when he died in 1727 he left a considerable fortune, much of it in shares. But not all his investments had been successful. The popular belief that shares in the South Sea Company would increase in value had been based on nothing more than a rumour; consequently the South Sea Bubble was destined to burst, and it duly did so, causing Newton financial pain. He is reported to have said that, for all his mathematical abilities, he 'could not calculate the madness of the people'.

Throughout the eighteenth and nineteenth centuries the fledgling financial markets of Newton's day grew into great birds of prey, offering countless opportunities for speculation. In addition to stocks and shares, people could speculate on annuities, insurance, and foreign exchange, to say nothing of lotteries. But the study of financial operations by mathematical methods made few advances, in spite of the steady progress in the study of probability and statistics. Even the use of pictures to represent financial data had made little progress after its introduction by William Playfair in 1786. Things began to change in the 1850s, partly as a result of the influence of Florence Nightingale.[1] Her work on improving the medical support for soldiers was firmly based on statistics and, like Playfair, she believed that data are more likely to be understood when presented pictorially. The same conviction led to a discussion on the subject at an International Conference of Statisticians held in 1857, and the idea was taken up by some of the leading theorists. One of them was Stanley Jevons, who conceived the idea that the observed 11-year cycle of sunspot activity had an effect on trade, because the consequent variations in the weather affected the production of crops. Jevons constructed diagrams that clearly showed periodic fluctuations of economic activity, now known as 'business cycles', but his sunspot theory did not catch on.[2]

By the end of the nineteenth century it was accepted that such 'charts' were an excellent means of representing the movements of prices over time. The chart shown in Figure 10.1 is taken from George Clare's *Money Market Primer and Key to the Exchanges*, published in 1893. It shows the fluctuations in two versions of the exchange rate between London and Paris for the year 1888.

Clare's diagrams show a marked degree of irregularity, even though they have been smoothed by taking weekly averages. Nowadays the rates change very frequently, and if the time-period was one hour rather than one year, the diagrams would look much the same. In the finance industry such diagrams are still known as *charts*, and modern-day numerologists devote their lives to studying them, in the hope of making money.

Of course, they are wasting their time—and our money, as the charges we pay for financial services provide their wages and bonuses. That does not mean that mathematics cannot be usefully employed to describe the movements of prices over time, provided that we do not pretend that it can predict them reliably. The 'madness of the people' remains the major factor that causes changes in the price of a financial asset, and that is much harder to predict than factors described by the laws of physics. We still have no reliable method of predicting earthquakes, so it should not come as a surprise to find that we cannot predict financial disasters.

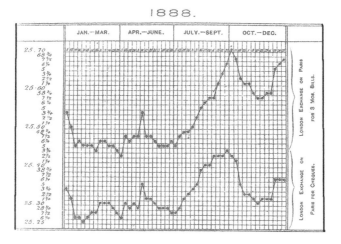

Figure 10.1 A financial chart from the end of the nineteenth century.

A model of how prices change

The first person to attempt to make a mathematical model of the movements of asset prices was a young French mathematician, Louis Bachelier. In 1900 he presented his doctoral thesis entitled *Théorie de la Spéculation*, in which he laid the foundations for the subject that is now known as financial mathematics. His work was truly remarkable, because he was able to work out many of the essential features of the problem, even though the rigorous mathematical foundations were not sorted out until later.[3] One of the fundamental problems was that the theory of probability had to be upgraded. The discrete models that suffice to deal with problems of throwing dice are not enough for situations where the variables can change over arbitrarily small intervals of time, and the methods of calculus have to be used.

Bachelier's model has since been amended and extended, but the basic ideas can still be explained in a way that Leibniz would have understood. The aim is to describe how the price S of an asset varies as a function of time t. In a small time dt, the change in S is dS, and as the numerical value of S is a purely arbitrary measure, it makes sense to consider the ratio dS/S. The now-standard model assumes that dS/S is the sum of two contributions, one deterministic and one random. The first one is an underlying trend, which contributes a change proportional to the time-interval, say $A \, dt$. The second one is a contribution resulting from the 'madness of the people', which we denote by $B \, dx$. It is assumed that the quantities A and B are characteristics of the asset, and they can be calculated from data on how the price has changed in the past. So this model leads to the equation

$$\frac{dS}{S} = A \, dt + B \, dx.$$

Based on this equation we can construct an imaginary chart of the price of an asset. Assume that the parameters have been estimated as $A = 0.01, B = 0.2$, and consider a period of 100 days, taking each dt as one day. Then the changes in S depend on 100 values of dx, one for each day; for example, they might be as given in the following table.

Day:	1	2	3	4	$\cdots$	100
dx:	+0.04	−0.06	−0.01	+0.02	$\cdots$	−0.03

Given the initial price S_0, it is easy to calculate the values at the end of each day, $S_1, S_2, S_3, \ldots, S_{100}$. The value of S_1 is $S_0 + dS$, and the rule says that

$$\frac{dS}{S_0} = (0.01 \times 1) + (0.2 \times 0.04) = 0.018, \text{ so } S_1 = S_0 + dS = 1.018 S_0.$$

Continuing in this way we can draw the chart, as in Figure 10.2, and work out the final price S_{100}. If we had been given a different sequence of dxs, we would get a different chart and a different final price S_{100}.

Bachelier's idea was to assume that the rule for choosing the dxs is a probabilistic one. For example, one very simple rule would be to allow dx to take only two values, say $dx = +0.02$ with probability p, and $dx = −0.02$ with probability $1 − p$. The point is that if the probability distribution of dx is known, then it is possible to work out the probability distribution of the final price S_{100}. This allows us to answer the questions that a speculator will ask, such as: What is the probability that the price will increase by at least 10 per cent?

Realistically, the behaviour of dx is unlikely to be determined by a simple rule like the one suggested above. A more convincing model requires a better rule, taking account of the fact that the time intervals dt must be allowed to vary, and become arbitrarily small. A subtle blend of probability and calculus is needed to set up a proper model of this situation. That was not available in Bachelier's day, but nevertheless he was able to obtain some significant results. His most important insight

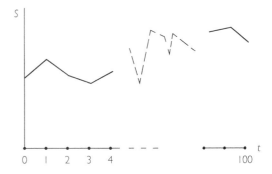

Figure 10.2 An imaginary chart of price movements over 100 days.

was that there must be some constraint on dx as dt approaches zero, otherwise the entire process will become trivial. As dx is the result of many independent actions by investors and speculators, Bachelier assumed that it follows a normal distribution. On this basis, he found that he could set up a satisfactory model, with the expectation of dx being zero and the expectation of $(dx)^2$ being proportional to dt.

Remarkably, a few years later the same conclusion was reached in another paper, written by a mathematician working in a quite different context. That mathematician was Albert Einstein, and the relevant paper was one of four that were published in his 'miraculous year' 1905. Among these papers we find the foundations of the theory of special relativity and the famous equation $E = mc^2$. But the relevant paper for us is concerned with a phenomenon known as Brownian Motion. In 1827 Robert Brown had observed that when particles of pollen are suspended in water they jump about in a curious way. Einstein believed that this motion provided evidence for the physical existence of molecules of water, the motion being caused by the bombardment of the pollen particles by the molecules. In attempting to set up a model, Einstein independently came to the same conclusion as Bachelier, that the displacement of a particle must be proportional to the square root of the time-interval in which it is measured.

As far as we know, Einstein was unaware of Bachelier's thesis, and indeed it remained unnoticed for many years. In that time there were several theoretical advances that justified the insights of both Bachelier and Einstein. In the 1920s the theory required to prove the existence of a mathematical model in which dx is normally distributed was developed by Norbert Wiener, and in the 1930s Andrey Kolmogorov strengthened the foundations of probability theory. Finally, from about 1944 onwards Kiyosi Itô showed how equations involving differentials of non-deterministic processes can be solved. These advances led to the subject that we now call stochastic analysis. It is still a very active area of research in the twenty-first century, partly because Bachelier's model of asset prices has become a fundamental tool in the real world of finance.

Forms of money

Throughout this book we have encountered some of the problems associated with the use of coined money as a means of exchange. These practical problems inevitably led to mathematical ones. The use of gold and silver coins was one of the main reasons why the algorithms of elementary arithmetic were developed, and eventually cast in a form that could be taught to young children. More subtle calculations were also required, such as those needed to measure the amount of precious metal in a coin, but they were taught only to a chosen few.

When large transactions were required, the use of coined money was not always convenient. In Europe this led to the Bill of Exchange (see Figure 5.2) and the need for calculations based on a variable exchange rate. In China the problem of large transactions was made worse by the only coin, the bronze 'cash', having low value.

By about 1100 AD payments involving large numbers of cash had become common, and the Chinese began to use paper notes, equivalent in value to 1000 cash. Similar developments elsewhere led to the growth of institutions that we should now call banks, although they were often engaged in a range of other commercial activities.

The use of paper notes gradually spread to the West, and they became a convenient means of making payments of all kinds.[4] The underlying principle was that Mr Affluent, who had deposited items at a bank (usually precious metals in some form), would be given notes that could be used to pay for goods. If Mrs Baker or Mr Butcher received such a note they could use it in the same way or, if necessary, they could present it to the bank of issue and be given the equivalent amount in gold or silver. Of course, the original idea was that a bank would only issue notes to the value of the items that it held on deposit, but this rule was often breached, mainly to facilitate trade by the granting of credit. Consequently, in the nineteenth century many private banks failed when it became clear that they could not redeem their notes for hard cash. For such reasons banknotes were often regarded with suspicion by the public, although they were an important tool in the financial policies of governments. The Bank of England had been founded in 1694 and, like the private banks, it issued banknotes. In 1797, when it was feared that the French might invade England, the Bank was forbidden to release its stocks of gold, and banknotes were offered instead. A contemporary print[5] shows the Prime Minister, William Pitt, handing banknotes to John Bull, a befuddled Englishman, while large bags of gold are hidden under the counter. Pitt's political opponents are whispering to John Bull, suggesting that he would find gold more effective than banknotes, if he were mugged by a fiery Frenchman. In the twenty-first century the issue of banknotes is still regarded as a convenient mechanism for manipulating the national economy, and we have become familiar with the term 'quantitative easing', a euphemism for printing more banknotes to avoid economic depression.

The use of paper notes was just one step along the path towards a more nebulous and abstract notion of money. The Romans had discovered that coins did not necessarily have to contain precious metal; they could be made to perform their function by decree, backed by force if necessary. Nowadays the developed world is accustomed to using token coins, banknotes, and other forms of paper money such as cheques, bonds, and share certificates. But these are by no means the most nebulous objects that are traded in the money markets of the twenty-first century. There is, for example, the *option*. In its simplest form, an option gives a speculator the right to buy an asset at a fixed price E at some future time T. If, at that time, the market price of the asset S_T is greater than E, then the speculator will take up the option and gain an amount $S_T - E$; otherwise she will not do so, and will gain nothing. The option can be traded at any intermediate time t, and its value at that time, V_t, clearly depends in some way on the asset price S_t. The obvious question is: If the probability distribution of S_t evolves according to a model like Bachelier's, what is the probability distribution of V_t?

To answer this question it is necessary to make some assumptions about the workings of the financial markets. By the 1960s knowledge of Bachelier's model had gradually filtered into mathematical circles, but it had found no practical application in finance. In particular, its relevance to the valuation of options had not been seriously discussed. One of the first people to spot this gap was the leading economist of the time, Paul Samuelson, who analyzed the basis of Bachelier's assumptions in terms of the economic theory of financial operations. But it was some years before his ideas were translated into a form that had direct practical application.

The key is the No-Arbitrage Principle, sometimes paraphrased as 'there is no free lunch'. It must not be possible for a trader to make a guaranteed profit simply by switching from one asset to another, however rapidly the transactions are made. In particular, the holder of an option should not be able to make a profit by exchanging it with the asset on which it is based. In 1973 Fischer Black and Myron Scholes saw how to combine this principle with Bachelier's model, and using the tools of stochastic analysis they obtained an equation for the relationship between the asset price S and the value V of an option based on that asset. The theory was extended later that year by Robert Merton, a pupil of Samuelson. One attractive feature of the Black–Scholes–Merton equation is that it has a close relation to the equation governing the flow of heat, which had been studied by mathematicians and physicists for almost 200 years.

The availability of tractable models led to a spectacular boom in the trading of options of all kinds. The Black–Scholes–Merton equation provided an explicit formula for option-pricing, which was a powerful, but misleading, incentive. The apparent ease of applying 'the formula' concealed the fact that it is based upon a mathematical model, which inevitably over-simplifies the real situation. Sadly, many of the people responsible for this boom understood very little about the mathematics or the assumptions on which it is based. Even if they were aware of the problems, they believed that the operation of the financial markets would automatically correct unsound decisions. The trading in options became increasingly bizarre, and the baroque convolutions of this activity reached heights that were, literally, fantastic: the alleged value of the options being traded was many times the value of all the world's real resources. It was the South Sea Bubble again, and again it burst.

Money, information, and uncertainty

The world of modern banking has another feature that tends to propel it into the realms of fantasy. Vast sums of money can be transferred in an instant, far more quickly than it has taken you to read this sentence. At the humble level of personal banking this affects us all directly, because the traditional methods of payment by cash or by cheque are gradually being displaced by payment using plastic cards. When we buy things at a shop or online, we use our cards to pay for them, so the

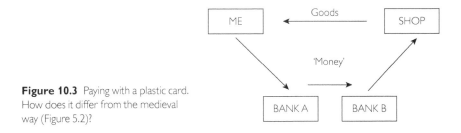

Figure 10.3 Paying with a plastic card. How does it differ from the medieval way (Figure 5.2)?

cards themselves can be thought of as a form of money. But it is worth looking more closely at what actually happens.

Our cards have the power to initiate a sequence of accounting operations (Figure 10.3). Suppose I use my card to buy a present for my wife costing 100 pounds. First, a message is sent to my bank telling them to deduct 100 pounds from my account: in other words the bank now owes me 100 pounds less than it did before. At the same time a message is sent from my bank to the seller's bank, transferring the 100 pounds to the latter, and finally, the seller's bank adds this amount to the seller's account. These operations take place instantaneously and at negligible cost. The result is that my bank now owes me 100 pounds less, and another bank owes the seller 100 pounds more. The account between the banks has been adjusted accordingly, so that their position is essentially unchanged. Significantly, the only physical changes that have taken place are in the records kept by the various parties. In this case money is represented by some numbers stored in computers: it is simply information. The information indicates how much money is owed by the banks to their customers, and my card allows me to alter this information in a specific way.

The situation described above is only one aspect of the relationship between money and information. Indeed, it is becoming clear that the concept of information is all-pervasive, even playing a part in the laws of physics. This viewpoint arises partly because we now have a satisfactory way of measuring information, which can be applied to the many ways in which information is expressed.

In the distant past humans began to exchange information by making signs and noises. Gradually the signs and noises became language, first spoken and then written. The process was stimulated by the requirements of exchange and trade, and as trade became more complicated, so more information was needed. Equally, the human contacts involved in trade were instrumental in transmitting information about other aspects of life, cultural, social, and even mathematical. Natural languages, written in a uniform way, were the means by which this information was disseminated.

A message written in a natural language can be regarded as a stream of symbols, such as those printed on this page. That was the viewpoint of the ninth-century Arab cryptographers when they used the method of frequency analysis to decipher coded messages. In any specific language, such as modern English, the stream of symbols has certain statistical properties, and these remain fairly constant across a wide range of sources, from *Hamlet* to *Harry Potter*. The simplest properties are the frequencies of the individual symbols, but there are also more complex properties involving relationships between successive symbols, such as the rule that q is usually followed by u. Consequently, a natural language has a certain redundancy: not all the symbols in a message are needed to convey its meaning. We use this property when we send text messages; for example, we might use the transformation

$$\text{Are you going to be late?} \longrightarrow \text{r u goin 2 b l8?}$$

Here 25 symbols (counting the spaces) have been replaced by 16, using the irregular encoding procedure known as 'textspeak'. We shall see that there are formal algorithms for such procedures, but already it is clear that to measure the amount of information in a message, we must look beyond the size of the raw data that it contains. In the case of natural languages the situation is quite complicated (although some good definitions can be made), but the basic ideas can be illustrated by looking at a simpler set-up.

Suppose there are just two possible symbols in the stream, and the only rule is that their relative frequency is fixed. For example, if we toss a coin repeatedly we shall obtain a sequence of Heads (represented by H) and Tails (represented by T). That was the situation studied by Jakob Bernoulli and Abraham de Moivre around 1700: they showed that if the coin is tossed many times, then we can make sensible estimates of the likelihood that the numbers of heads and tails lie within a given range (Chapter 8). If the coin is 'fair' the stream of data produced by this process might look like this:

TTHTHHTHTTTHHTHTHHTHTTTHTHHTTHTTHHHTTTHHTHTHHTTH.

On the other hand, if the data are generated by a process in which two outcomes have different probabilities, then the stream may look rather different. For example, we might repeatedly throw a pair of dice, so that there are $6 \times 6 = 36$ equally probable outcomes, as shown in Figure 10.4.

Suppose we record W (for White) when the two dice show the same number, and G (for Grey) when they do not. Then 6 of the 36 outcomes are W, and the remaining 30 are G. It follows that we shall produce a stream of Ws and Gs in which the probability of White has the value $6/36 = 1/6$, and the probability of Grey has the value $30/36 = 5/6$. Now the stream of data might look like this:

GGGGWGGGGGGWGWGGGGGGGGGWGGGWGGGGGGWWGGGGGGGGG.

Which of these two streams, the H-T or the W-G, provides more information? The correct answer is the H-T stream. To understand why, consider the uncertainty of someone who is receiving the data, symbol by symbol. In the first case each new symbol provides the answer to a question in which the two outcomes (H or T) are equally probable. But in the second case the outcome W is significantly less likely than G, which means that the uncertainty is less. Thus, from the viewpoint of

1 1	1 2	1 3	1 4	1 5	1 6
2 1	2 2	2 3	2 4	2 5	2 6
3 1	3 2	3 3	3 4	3 5	3 6
4 1	4 2	4 3	4 4	4 5	4 6
5 1	5 2	5 3	5 4	5 5	5 6
6 1	6 2	6 3	6 4	6 5	6 6

Figure 10.4 The 36 equally probable outcomes when two fair dice are thrown.

the receiver, the information provided by each new symbol is also less. This basic insight lies at the core of the theory: to measure the information contained in a stream of data, we measure the amount by which our uncertainty is reduced when the data are provided.

Measuring information

The formulation of a mathematical definition that captures the relationship between information and uncertainty is due to an American engineer and mathematician, Claude Shannon. His interest in the subject began in 1937, when he wrote a thesis for his master's degree describing how electrical circuits can be designed to solve logical problems. Shannon's research was the catalyst for major technical advances in engineering, but fortunately the underlying logical framework can be explained without going into the practical details.

As an example, consider the following problem. You choose a whole number k in the range 1 to 8, and I wish to find your number by asking questions with a Yes/No answer. One way to do this would be to begin by asking 'is $k = 1$?'. If the answer is No, then I would ask 'is $k = 2$?', and so on. In the worst case, if the number you have chosen is 8, I shall have to ask eight questions using this strategy. More generally, if all your choices are equally likely, then I would expect to ask at least four questions.[6] However, there is a better strategy. The idea is to split the possibilities into two parts, and discover which one of them contains the answer. In this case the first question would be 'is k in the range 1 to 4?'. Knowing the answer to this question reduces the number of possibilities from eight to four. A second question

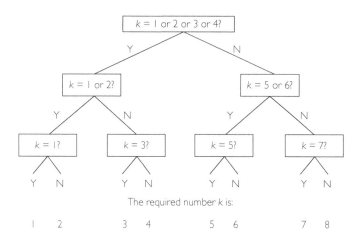

Figure 10.5 How to find a number in the range 1 to 8 by asking three questions.

reduces the possibilities to two, and a third question reduces the possibilities to one (Figure 10.5). Thus I can find your chosen number by asking only three questions.

If there are n alternatives instead of 8, how many Yes/No questions will be needed? The method illustrated in the diagram shows that one question will resolve two possibilities, two questions will resolve four possibilities, three questions will resolve eight possibilities, and generally x questions will resolve (up to) 2^y possibilities. Putting it another way, to discover one of n alternatives, we need x questions, where x is (approximately) the number such that $2^x = n$. The inverse relationship between exponents and logarithms (Chapter 7) suggests that we call this number the logarithm[7] of n to the base 2. We shall write it as $\log_2 n$.

The foregoing analysis suggests that we should consider $\log_2 n$ as a measure of the uncertainty when n alternatives are equally likely. This is the same as saying that $\log_2 n$ measures the amount of information communicated when one of the n alternatives is revealed. Shannon used this idea in a famous paper entitled 'A Mathematical Theory of Communication', published in 1948. The central concept was a definition of uncertainty, which covered situations where events can occur with differing probabilities. That kind of situation was discussed in the previous section, when we considered two sources, each producing a stream of symbols. In both cases there were two alternatives for each symbol, Heads/Tails or White/Grey. If the two alternatives occur with probabilities p and q, where $q = 1 - p$, Shannon's measure of uncertainty is

$$p \log_2 \left(\frac{1}{p} \right) + q \log_2 \left(\frac{1}{q} \right).$$

Equivalently, this is the amount of information conveyed by each symbol in the stream.

The unit of information is called a *bit*, derived from the words 'binary digit' used to describe the symbols 0 and 1. To decide between these two alternatives one Yes/No question is required, so this represents one bit of information. Rather confusingly, the symbols 0 and 1 themselves are also referred to as bits (of data).

Shannon's general definition of information (or uncertainty) is a simple generalization of the formula displayed above.[8] It turns out that if we compile a list of properties that we would normally associate with the notion of uncertainty, then there is essentially only one definition that fits all of them, and that is Shannon's. In fact, the definition is justified, not only because it has the right properties, but also because it can be used to formulate and prove significant results. As an example, let us return to the two streams of data discussed in the previous section, the Heads-Tails stream and the White-Grey stream. In the case of the H-T stream p and q are both 1/2, so $1/p$ and $1/q$ are both 2. The logarithm (to base 2) of 2 is 1, so the formula gives the value 1. This means that each symbol, H or T, conveys one bit of information. If the coin-tossing experiment produces a stream of 1000 symbols, then 1000 bits of data are needed to represent the outcome. On the other hand, in the case of the stream of Ws and Gs, the probabilities are $p = 1/6$ and $q = 5/6$. Putting these values in the formula gives the result 0.65 approximately, which means that each symbol provides only about 0.65 bits of information. So, if the theory is correct, it should be possible to represent a stream of 1000 Ws and Gs using fewer than 1000 bits of data (hopefully about 650). The technique for doing this is known as *data compression*.

The idea is to split the stream into blocks of two symbols, so that each block is WW, WG, GW, or GG. The properties of the source imply that these blocks occur with the following probabilities:

WW	WG	GW	GG
1/36	5/36	5/36	25/36.

We now represent the blocks in a simple way, using the symbols 0 and 1. Specifically, we encode the data using the following 'codewords' for the four blocks: WW → 111, WG → 110, GW → 10, GG → 0. The crucial point is that the most likely block GG is represented by a codeword with just one symbol; it is true that some other blocks are represented by longer codewords, but they occur less frequently, and so on average the stream of data will be shortened. In fact, the codeword of length 1 is used with probability 25/36, the codeword of length 2 is used with probability 5/36, and the codewords of length 3 are used with probability $5/36 + 1/36 = 1/6$, so the average number of bits required to represent a block of two symbols is

$$\left(1 \times \frac{25}{36}\right) + \left(2 \times \frac{5}{36}\right) + \left(3 \times \frac{1}{6}\right) = \frac{53}{36}.$$

A stream of 1000 symbols contains 500 blocks, so we expect the number of bits used in the encoded form to be about 500 $\times$ 53/36, which is approximately 736. This is a significant saving. We could use better ways of representing the data, starting with longer blocks, and thus approach more closely the theoretical minimum of 650 bits.

It only remains to explain why no information has been lost: in other words, why the original data can be recovered, using only the encoded stream of 0s and 1s. At first sight, this looks rather unlikely: for example, how would you decode the following stream?

$$01000100011001000001110001000011000010100010$$

It must be stressed that you can only use the rule for encoding the blocks, as given above; there are no additional hints in the form of punctuation. The reason why decoding can be done is that we have chosen our coding rule with a very special property: no codeword is the initial part of another. This enables us to decode the stream of 0s and 1s in the simplest possible way, by reading off the corresponding blocks in order. The first symbol is 0, which stands for GG, and as there is no other codeword beginning with 0, the first block must indeed be GG. Next we have 1, which is not a codeword, so we continue until we have a complete codeword, in this case 10, representing GW. The construction ensures that 10 is not the beginning of another codeword. The principle is most easily grasped by working through an example, such as the one given above: you should find that these 43 bits of information represent a stream of 64 Ws and Gs.[9]

This brief outline cannot do justice to Shannon's pioneering work. His discoveries provide the basis for stating theorems about such things as the capacity of an information channel, and for proving those theorems with the full force of mathematical precision. The results can be used to predict how systems will behave, and they have been successfully applied to a wide range of problems in the telecommunications industry. This situation should be compared with the relative lack of success in the case of modern financial theories, as discussed earlier in this chapter. The difference may occur because information theory is applied to situations in which the behaviour is governed by the laws of physics, whereas option-pricing theory is applied to situations in which human behaviour is a significant factor.

CHAPTER 11

Can Mathematics Keep Us Safe?

The information that you supply before you send mail electronically is a kind of key, designed to provide you with a modest level of security. Better security requires more sophisticated methods, and these are now based on algorithms that utilize the mathematical properties of prime numbers. Unfortunately, there are still some unanswered questions about the systems that are currently in use. So if you thought this was the end of the story. . . .

The search for security

Money, in the form of coins and jewels, was traditionally kept under lock and key. Wealthy medieval families would use a strong box with a large key, both of which were carefully concealed. Later, the box might be kept in the vaults of a bank, behind securely locked doors. In both cases a potential thief might be aware of the location of the box, but to steal the money he would need to find the keys. An analogous principle was applied to the sending of secret messages for military and diplomatic purposes: the means of communication might be easily discovered, but the 'keys' had to be kept secret. The result was a long-running battle in which the code-makers tried to make better keys, and the code-breakers sought better ways of finding them. The code-makers were winning in Julius Caesar's time, but by the ninth century the Arab code-breakers had discovered how to find the key to systems like Caesar's, using the method of frequency analysis (Chapter 5). Gradually the code-makers regained the ascendancy and, after the invention of the Vigenère system in the sixteenth century, the system was thought to be unbreakable. But in the nineteenth century improvements in statistical techniques led to effective methods of finding a Vigenère key, and the battle between the code-makers and the code-breakers was rekindled. In the first half of the twentieth century the emphasis shifted towards systems based on machines that could scramble data in very complicated ways, so that the number of possible keys was astronomically large.[1]

Nowadays cryptography is a routine part of our everyday lives, although we may not always be aware of it. But, as always, we are acutely aware of the need to keep our money secure. Because a great deal of our money is not kept in a tangible form, but in the form of information, the problem of keeping our money safe and the problem of keeping our messages secret have become almost identical. The messages initiated by our plastic cards must be sent and received safely: if all the parties involved are to be kept happy, the entire operation must be carried out with a high level of confidentiality. So we insist on several practical safeguards, such as *authentication* (instructions that purport to come from me must really do so), *integrity* (no-one else should be able to alter the instructions in the course of transmission), and *non-repudiation* (I must not be able to claim that I did not issue the instructions in the first place). The attempts of our governments to construct a legal framework for these matters are, as yet, rather primitive, witness the curious situation in the UK of having both a Data Protection Act and a Freedom of Information Act.

At a more practical level, it is clear that the basic tools of the code-makers have been introduced into our financial affairs. My plastic card has a 16-digit number on the front, and another shorter number on the back, and it contains a 'chip' that can do some mysterious operations with these numbers. I also have a 'pin' (Personal Identification Number) which I must memorize and supply whenever I use my card. These numbers form a kind of cryptographic key. But, as we shall now explain, the most sophisticated modern cryptosystems differ from the traditional ones in the way that the keys are used.

A feature of all the traditional systems was that a single key was used by the sender to encrypt the message and by the receiver to decrypt it. The problem with this procedure was that a separate and secure method for communicating the key was required, before the parties could begin to use the system. Cryptographers referred to this as the Key Distribution Problem, and it led to great practical difficulties, especially in times of war. A radical new approach was developed in the 1970s, based on a different way of using keys. The fundamental idea is that a typical user, let us call him Bob, has two keys, a 'public key' and a 'private key'. The public key is used to encrypt messages that other people wish to send to Bob, and the private key is used by Bob to decrypt these messages. The security of the system depends on ensuring that Bob's private key cannot be found easily, even though everyone knows his public key. In the rest of this chapter we shall look at the mathematical principles underlying this *public-key cryptography*.

Some things are easy, but some are not

Our reliance on computers tends to make us forget that, originally, arithmetic was a difficult subject. The Egyptian and Babylonian scribes were highly trained, but the practice of their art also required an innate facility for handling numbers. In the course of many centuries the algorithms improved significantly, and eventually writers like Edmund Wingate in the seventeenth century could hope to teach them to unwilling schoolboys. More recently the same algorithms have been entrusted to our computers, which work much faster than the schoolboys, and make less noise.

The electronic revolution has led to several new insights into the process of computation itself. In the 1930s several mathematicians began to consider how computation might be described in mathematical terms. They were motivated partly by logical problems about the limits of computation, and partly by the increasing use of electro-mechanical machines to do long and complex calculations. The most successful model of computation was constructed by Alan Turing (1912–54). His mathematical description of what we now call a Turing Machine has turned out to be of fundamental importance, and in the 1940s the development of electronic computers was guided to some extent by his work.

To illustrate the current state of the art, suppose I ask my computer to find the result of multiplying two numbers,

$$r = 1907180854589209641162363757488357797106749590673031653701$$
$$6839226001220767984427385832966637999862924555 1661101,$$

$$s = 3390914846854132702461119005170659611590183264349349314$$
$$1183991491544989934434385226970516014449975860700436 41531.$$

It gives me the answer almost instantaneously:

$$n = 6467087875462503741036180030918781080605161336528514858 51$$
$$51943834972903712651341316315961925108940083589520724084 51$$
$$48906831474789526624665917802693899847281141689594634884 00$$
$$95581638293010126348584409544381895712500040785631.$$

That such large numbers can be multiplied so quickly, using standard notation and algorithms, justifies the claim that this kind of arithmetic is now easy. However, in the 1960s people began to discover that not all computational problems share this property, and a theory of Computational Complexity was developed to try to explain this observation. The first step was to understand why some algorithms, such as the method of long multiplication that we teach in our schools, work just as well for large numbers as for small ones. The reason is that numbers can be represented in a compact notation. The result of multiplying 2 by itself 50 times is huge, but in 1633 Nicholas Hunt wrote it down using only 16 digits (Figure 7.3). For the purpose of calculation, it is the size of the representation that matters, not the size of the number itself. With decimal place-value notation n digits are enough to represent any number less than 10^n, and we can carry out the operations of elementary arithmetic efficiently using this notation. We have seen how easy it is for a modern computer to multiply two large numbers, each with over 100 digits; in fact the computer does this using essentially the same method that was used by ibn Labbān to multiply two three-digit numbers a thousand years ago (Figure 4.5). The basic steps are single-digit multiplications: each digit of the first number is multiplied by each digit of the second number. So ibn Labbān needed $3 \times 3 = 9$ steps, and if the two numbers have n digits, then we need $n \times n = n^2$ steps in all.

Generally, to measure the efficiency of an algorithm, we use the method described in the previous paragraph. That is, we look at the relationship between the size of the input and the number of steps required. For the latter an estimate is all that is needed, because in practice we are only interested in estimating roughly how long the algorithm will take. Suppose we have access to a computer that can perform one billion (10^9) steps every second, a not unreasonable assumption. Then the time required to multiply two 20 digit numbers is $20^2/10^9$ seconds, or 0.0000004 seconds. Of course, as n increases the time also increases, but only at a modest rate. If n is 100, the time required is only 0.00001 seconds. Even if we have a problem that needs a more complicated algorithm, requiring n^3 steps (say), the results are much the same. The computation takes a little longer, but it is still feasible because the time needed can still be measured in fractions of a second, as in the following table.

Input size:	$n = 20$	$n = 40$	$n = 60$	$n = 100$
n^2 steps:	0.0000004	0.0000016	0.0000036	0.00001
n^3 steps:	0.0000008	0.0000064	0.0000216	0.001.

The results given above justify the use of the word 'easy' for problems that can be solved by an algorithm for which the number of steps required is n^2 or n^3, or generally any fixed power n^m. Such algorithms are said to work in *polynomial-time* and the corresponding problems belong to a class that we call P. There are polynomial-time algorithms for carrying out the basic operations of elementary arithmetic, so these operations belong to the class P, and they are relatively easy.

It turns out that many other computational problems also belong to the class P. This means that such problems can be solved by using efficient algorithms (expressed in a computer programming language), even when the numbers involved have hundreds of digits. However, it is essential that an efficient algorithm is used. For example, recall the words of Mersenne, written in the seventeenth century. 'In order to decide whether given numbers consisting of 15 or 20 digits are prime or not, ... not even a whole lifetime is sufficient to examine this, by any method so far known'.

Mersenne's statement was correct, because he was referring to the algorithm that checks whether a number is prime by trying to find factors of it. Indeed, it is still (2015) believed that finding the factors of a number is not an easy problem. On the other hand, Mersenne was careful to allow for the possibility that there might be a different algorithm for the primality problem, one that does work in polynomial time. It was not until 2002 that such an algorithm was found, by three young Indian mathematicians, Manindra Agrawal, Neeraj Kayal, and Nitin Saxena.[2]

The algorithm of Agrawal, Kayal, and Saxena shows that testing for primes is in the class P, by a very ingenious theoretical method. In practice a different method, introduced in the 1980s is used; it is not strictly a polynomial-time algorithm, but it is nevertheless effective. This algorithm is based on two ideas that go back to Fermat: his Little Theorem (Chapter 6), and the laws of probability that he developed in conjunction with Pascal (Chapter 8). The idea is quite simple. If we wish to test whether a given number n is prime, we choose any number x between 2 and $n - 1$ and ask the Fermat Question:

$$\text{Is } x^{n-1} \text{ congruent to 1 mod } n?$$

The calculation of x^{n-1} modulo n can be done in polynomial time, so the test is easy to apply. If the answer is No, we can conclude that n is not prime, because otherwise Fermat's Little Theorem would be contradicted. But if the answer Yes, n may or may not be prime, and (at first sight) we are stuck. In 1976 Gary Miller suggested a way of overcoming this problem with the Fermat Question, and in 1980 Michael Rabin proved a theorem that makes it practicable. The Miller–Rabin Question is slightly more complicated than the Fermat Question, but the calculation can be done in polynomial time. The significant point is that the new question has the following property: when n is not prime the answer Yes will occur for not more than 25 per cent of the possible values of x. So if we choose x at random, the probability that we get a false-positive answer is 1/4 at most. And if we repeat the test k times, each with a new random value of x, the probability of getting k false positives is $(1/4)^k$. For example, if we do it twenty times, the probability of a mistake is much less than one in a trillion (10^{12}). The Miller–Rabin Test is now commonly used to find primes with up to 1000 digits, and there are no known instances of it failing.

Although problems like primality testing have now been shown to be easy, not all problems fall into this category. This situation plays a crucial part in modern cryptography, as we shall explain in detail in the next section. For the moment, let us consider a simple example. Suppose Eve is an identity-thief, who is trying to obtain the password for my bank account. I have chosen a password which is stored on a computer in binary form, something like 01100 11101 . . . 01010 10111. If the password has n bits, then there are 2^n possibilities; each one of them must be tested. Here the fact that the number of steps is 2^n has a remarkable effect on the time taken to carry out the search. For example, suppose Eve is using a computer that performs one billion steps every second. With a 20-bit password the number of steps is 2^{20}, which is about one million, so the total time required is only one-thousandth of a second. Doubling the size of the input from 20 to 40 means that the number of steps increases from 20 to 40, that is $2^{20} \times 2^{20}$, or about one trillion, and now the time taken is about 1000 seconds, or 16 minutes and 40 seconds. At this level my password is clearly still not safe, but for larger values of n the effect on the time required is spectacular.

Input size n:	20	40	60
Time for 2^n steps:	0.001 seconds	16+ minutes	about 200 years

So if my password has 60 binary digits, it is safe. In fact, even if Eve gets a faster computer, that will not make a lot of difference to the general conclusion.

In the 1960s it was observed that there are some practical algorithms that seem to display the same behaviour as Eve's 'brute force' password search. Several of them were being used in operational research, the branch of mathematics that deals with the allocation and scheduling of resources, where the practitioners found that methods that work fairly well for small problems fail completely when the size of the problem increases. Another problem where similar difficulties occurred was the old one of finding the factors of whole numbers. Although multiplying two large numbers is easy, 'unmultiplying' them (that is, finding the factors of the product) is not. If the number is small, finding factors appears to be quite straightforward: for example it is fairly easy to show 'by hand' that the factors of 1001 are 7, 11, and 13. Even a number like $2^{67} - 1 = 147573952589676412927$, which had defeated Mersenne in 1643, had been factored by Frank Nelson Cole in 1903. He had no mechanical aids, but it was, as he said, 'the work of three years of Sundays'. The advent of the electronic computer seemed to offer the prospect of rather more instant success, and indeed I have programs on my computer that can do Cole's job very quickly. When I ask for the factors of $2^{67} - 1$ the answer appears immediately: $193707721 \times 761838257287$.

But no-one has been able to find an algorithm that will work for really large numbers, where, at the time of writing, 'really large' means over 300 decimal digits. It is this hiatus that is currently being exploited to ensure that our financial affairs remain secure.

How to keep a secret: the modern way

The significant fact that emerges from the discussion in the previous section is that there are some mathematical operations that are easy to do, but hard to undo. This observation throws new light on our metaphorical model of the cryptographic process in terms of locking a box with a key. In simplistic terms, if we have a key that locks the box easily, we assume that the same key will unlock the box just as easily. But now we have to recognize the possibility that unlocking with the same key can be hard, although there may be a different key that does work easily. This is the basis of public-key cryptography, proposed by Whitfield Diffie and Martin Hellman in 1976.

Diffie and Hellman described the central idea of public-key cryptography, but they did not set up a working system. That was done in the following year by Ronald Rivest, Adi Shamir, and Leonard Adleman, and their system, known as RSA, became famous almost overnight, mainly because it was described by Martin Gardner, in one of his monthly articles in the *Scientific American*.[3] It is based on the mathematical results described in the previous section, and it illustrates the basic principles of public-key cryptography very elegantly.[4]

RSA is a collection of several algorithms. The first is a procedure that enables a user (say Bob) to calculate two numerical keys. We call them his *private key* and

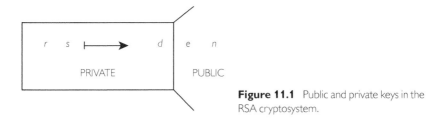

Figure 11.1 Public and private keys in the RSA cryptosystem.

his *public key*. Bob begins by choosing two numbers r and s, which must be prime numbers. As we now know, he can easily find such numbers with several hundred digits. Then he uses r and s to calculate three other numbers, denoted by n, d, and e. The number n is just r times s and, although the numbers are large, we know that it can be found easily by long multiplication. For the numbers d and e Bob must ensure that the product de is congruent to 1 modulo $(r - 1)(s - 1)$; this too can be done quite easily. The numbers n and e comprise Bob's public key, and he makes them available to everyone. But d is his private key, and he keeps this number secret, together with the prime numbers r and s that he used to defile d (Figure 11.1).

To complete the RSA system, two more algorithms are needed: one for encrypting messages and one for decrypting them (Figure 11.2). The input to the encryption algorithm is the original message, together with the public values of n and e. The input to the decryption algorithm is the encrypted message, together with the private value of d (and n). When someone (say Alice) wishes to send Bob a message, she uses his public key to encrypt it, and Bob uses his private key to decrypt it. As a consequence of the way in which the keys have been chosen, the algorithm for decryption is the inverse of the algorithm for encryption. In other words, the decrypted message is the same as the original one. The proof relies on some basic results discovered by Fermat and Euler: a salutary reminder that mathematics can be well ahead of its time, in practical terms.[5]

The effectiveness of RSA depends on two things. It is efficient, because the encryption and decryption algorithms used by Alice and Bob are easy, in the sense that they work in polynomial-time. On the other hand, it is believed to be secure, because no-one has found an easy way of decrypting the encrypted message without knowing Bob's private key. We do not have an 'easy' algorithm for calculating the private numbers r, s, and d, even though the public numbers n and e are known. Unfortunately, this is not a proven fact in mathematical terms.[6]

For this reason it is worth looking more closely at how RSA works in practice. Bob must begin by finding the two primes r and s. They must be large but, as we have observed, good practical algorithms for this task have been available since

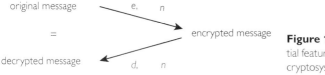

Figure 11.2 The essential feature of the RSA cryptosystem.

the 1970s. For example, if he wants to ensure that there are over 100 digits in both numbers, he could ask his computer for (say) a prime number greater than 13^{99} and a prime number greater than 19^{88}. Almost instantaneously, the computer will come up with the numbers r and s displayed earlier. Again, when Bob asks his computer to multiply r and s, the result, a number with over 200 digits, appears immediately. On the other hand, suppose Eve has intercepted an encrypted message from Alice to Bob. She knows Bob's public key, the numbers n and e, but this information does not allow her to carry out the decryption algorithm, because she cannot calculate Bob's private key d. If she could find the prime factors r and s the task would be trivial, because the rules for doing Bob's initial calculation are common knowledge. But Eve knows only n, not the factorization $n = rs$. She could ask her computer to find r and s, but if n is a number with several hundred digits the algorithms currently available will not respond to Eve's command, even if she is prepared to wait for a lifetime. When Martin Gardner first wrote about RSA in the *Scientific American*, he illustrated the strength of the system by challenging his readers to find the prime factors of a certain number with 129 digits. It took 17 years to solve this problem, using the combined efforts of over 600 people. Subsequently, cash prizes were offered for factoring other large numbers, and numbers with up to 232 digits were successfully factored. But (as far as we know) there was no breakthrough. No fundamentally new ideas were discovered, the successes being made as a result of minor improvements in strategy and the deployment of computing resources. The cash prizes were withdrawn in 2007, but some of the unconquered numbers can still be found on the internet, and doubtless there are many people working hard to factor them.[7] If an easy way of solving such problems were found, there would be serious consequences.

Can mathematics keep us safe?

The question asked at the start of this chapter deserves to be considered in a wider context, even if the answer is inconclusive. On the evolutionary scale of billions of years, the period covered in this book is but the blink of an eye. In a mere 5000 years mathematics has helped to transform the human condition, and it has become our best hope of understanding that condition. But we cannot overlook the fact that, in the wrong hands, mathematics can create enormous problems. In the next hundred years there will surely be progress (of the traditional kind), and mathematics will help to make it happen. On the other hand, humankind is now faced with the real possibility of extinction. Is mathematics a safeguard against extremism of all kinds, or is it a dangerous weapon? It would be good to end with a comforting reference to the lessons of history, but the lessons provide us only with hope, not certainty.

NOTES

CHAPTER 1 THE UNWRITTEN STORY

1. The best introduction to the mathematical aspects of our story is Jacqueline Stedall's *The History of Mathematics: A Very Short Introduction* (Oxford: Oxford University Press, 2012). Many older books on the history of mathematics approach the subject almost entirely from the viewpoint of a professional mathematician, and offer few insights into the context in which the fundamental ideas were developed. In recent years historians have tended to take a wider view, and this is exemplified in several contributions to the *Oxford Handbook of the History of Mathematics*, edited by Eleanor Robson and Jacqueline Stedall (Oxford: Oxford University Press, 2009).
2. The Prometheus myth is in Hesiod's *Theogony*. An online translation is available at www.sacred-texts.com/cla/hesiod/theogony.htm (accessed 1 June 2015).
3. The cut-and-choose procedure is discussed (without woolly mammoths) in: Steven J. Brams and Alan D. Taylor, *Fair Division* (Cambridge: Cambridge University Press, 1996).
4. A recent account of archaeological discoveries of tally-sticks and bones can be found in Stephen Chrysomalis's chapter 'The cognitive and cultural foundations of mathematics' in the *Oxford Handbook* (see note 1), pp. 495–518.
5. 'The Debate between Grain and Sheep' is available in the online *Electronic Text Corpus of Sumerian Literature*, etcsl.orinst.ox.ac.uk (accessed 1 June 2015).
6. For the origins of money, including the views of economists and anthropologists, see Glyn Davies, *A History of Money* (Cardiff: University of Wales Press, 2002). See also the thought-provoking account by David Graeber, *Debt: The First Five Thousand Years* (New York: Melville House, 2011).
7. Arguments against the idea that there was a global mathematical science in the third millennium BC can be found in Wilbur Knorr's article 'The geometer and the archeo-astronomers on the pre-historic origins of mathematics', *British Journal for the History of Science* 18 (1985) 197–212.

CHAPTER 2 THE DAWN OF CIVILIZATION

1. A good account of current thinking on ancient mathematics is contained in *The Mathematics of Egypt, Mesopotamia, China, India, and Islam* (Princeton: Princeton University Press, 2007), edited by Victor Katz. This book is referred to as [K] in the following notes.
2. Eleanor Robson's chapter on 'Mesopotamian Mathematics' [K, 58–182] is recommended; also her book *Mathematics in Ancient Iraq: A Social History* (Princeton: Princeton University Press, 2008).
3. Annette Imhausen's chapter on 'Egyptian Mathematics' [K, 7–56] contains a good account of the Rhind Papyrus.

4. An alternative expression for 2/9 is $1/5 + 1/45$. The convention was that the component fractions must decrease in size, so the expression $1/9 + 1/9$ was not allowed.

5. The tablet W.19408.76 is described by Robson [K, 73]. Following the upheavals of the Iraq War, its fate is a matter of some concern.

6. The tablet with the problem about a cylindrical sila-measure is Haddad 104, which was found in a controlled archaeological excavation at the town of Me-Turan [K, 123–125].

7. As 30 fingers make one cubit and 360 fingers make one rod, the number of cubic fingers in a volume-sar is $360 \times 360 \times 30$. This is the same as 21600×180.

8. An interesting collection of drawings showing the use of the equal-arm balance in Egypt was published by Hippolyte Ducros, 'Études sur les balances Égyptiennes, *Annales du Service des Antiquités de l'Égypt* 9 (1908) 32–52. Our Figure 2.12 is his no.38.

9. Most books on ancient metrology published before about 1970 are unreliable and cannot be recommended. A brief account of some of the confusion created by these books is given in Chapter 2 of Karl Petruso's *Ayia Irini: The Balance Weights* (Mainz: Verlag Philipp von Zabern, 1992).

10. This weight can be seen in the British Museum Online Collection: mina weight, number 91005.

11. The tablet referring to the exchange rate for grain and silver is discussed by Robson [K, 141]. See also the *Electronic Text Corpus of Sumerian Literature* (as note 5, Chapter 1) item 3.1.17.

12. Typical fragments of silver and gold can be seen in the British Museum Online Collection: finds from Amarna, number 68503.

13. As note 8, no.35.

CHAPTER 3 FROM TAX AND TRADE TO THEOREMS

1. A well-illustrated account of the evolution of coinage is *Money: A History* (London: British Museum Press, 1997), edited by Jonathan Williams.

2. Markus Asper's chapter on 'The two cultures of mathematics in ancient Greece' in *The Oxford Handbook of the History of Mathematics* (Oxford: Oxford University Press, 2009, pp. 107–132) is recommended.

3. This is one of many scenes depicted on the vase. It is on display in the Museo Archaeologico Nazionale, Naples.

4. The Greek alphabet was derived from the one invented by the Phoenicians, as was the Hebrew alphabet. The Hebrews also adopted the idea of representing numbers by alphabetical symbols.

5. Alain Scharlig, *Compter avec des calloux: le calcul elementaire sur l'abaque chez les ancienas Grecs* (Lausanne: Presses Polytechniques et Universitaires Romandes, 2001).

6. The Vindolanda tablets can be viewed online at vindolanda.csad.ox.ac.uk (accessed 1 June 2015).

7. S. Moorhead, A. Booth, and R. Bland, *The Frome Hoard* (London: British Museum Press, 2010).

8. A full discussion of the evidence about Pythagoras and his links with mathematics can be found in the online *Stanford Encyclopedia of Philosophy* (summer 2014 version), plato.stanford.edu (accessed 1 June 2015).

9. A detailed discussion of the manuscripts is given by David Fowler in *The Mathematics of Plato's Academy* (Oxford: Oxford University Press, 1999). A useful and accessible guide to what is currently accepted as the standard text has been prepared by David Joyce, and is

available at aleph0.clark.edu (accessed 1 June 2015). A critical edition, taking into account all the currently available materials (including Arabic manuscripts), is much needed.

10. There is an alternative proof in Euclid, Book VI, 31. It is shorter, but it depends on a longer chain of previously proved results.

11. In the proof of the Pythagorean Theorem, the second part is done 'similarly', using the triangles shown below.

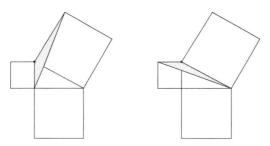

12. We cannot leave Archimedes without referring to the dramatic story of the discovery of a tenth-century copy of a previously unknown work by him. The story is described by Reviel Netz and William Noel in *The Archimedes Codex* (Cambridge, Mass.: Da Capo Press, 2007).

13. The proof in Euclid actually shows that given any three primes, another one can be found, but the same method will work for any given number of primes.

14. The proof that every whole number can be expressed uniquely as a product of primes is based on propositions that can be found in Euclid's Book VII, numbers 30, 31, 32. For a modern account see, for example, the author's *Discrete Mathematics* (Oxford: Oxford University Press, 2002).

CHAPTER 4 THE AGE OF ALGORITHMS

1. Many of the topics discussed in this chapter are described in more detail in Victor Katz (ed.), *The Mathematics of Egypt, Mesopotamia, China, India, and Islam* (Princeton: Princeton University Press, 2007). This book is referred to as [K] in the following notes.

2. For calculations with the Chinese rods see [K, 194–9]. Sometimes the rods were arranged in slightly different ways.

3. For the problems of the dust-board see [K, 532].

4. See E.G. Richards, *Mapping Time: The Calendar and its History* (Oxford: Oxford University Press, 1999).

5. British Library, Ms Royal BA XI f.33v.

6. Wingate's *Arithmetic* was first published in 1630 and remained in print for 130 years. The extract shown is from the 1735 edition, revised by John Kersey.

7. For al-Uqlīdisī, including a picture of the manuscript, see [K, 530–3].

8. For ibn Labbān's calculation, see [K, 533–5].

9. For the Bakshali Manuscript see [K, 435–41]. Sutra 27 contains a slightly more complicated version of the problem of calculating the average fineness of a set of coins. The answer to the simplified version given here is 80 per cent. In general, if the coins have weights $m_1, m_2, \ldots, m_n$ and finenesses $f_1, f_2, \ldots, f_n$, the answer is

$$\frac{m_1 f_1 + m_2 f_2 + \ldots + m_n f_n}{m_1 + m_2 + \ldots + m_n}.$$

10. Many images of the Cuerdale hoard are available in the British Museum Collection Online. It was deposited around 905, soon after the Viking settlers in England began minting coins in the Anglo-Saxon style, and some of these coins are found in the hoard. Coinage in the Scandinavian homelands of the Vikings did not begin until the end of the tenth century.

11. The life and work of Gerbert is described by N.M. Brown, *The Abacus and the Cross* (New York: Basic Books, 2010).

12. Several manuscripts illustrating Gerbert's abacus are now available online. One of them is St John's College Oxford, Manuscript 17, a collection of scholarly works made at Thorney Abbey in England around 1110. This manuscript can be viewed in its entirety, with full critical apparatus, at digital.library.mcgill.calms-17l (accessed 1 June 2015). A slightly later manuscript on the Bodleian Library website contains illustrations of the procedures for doing arithmetic, using the numbered counters: Ms. Bodleian Auct. F.1.9 f.65v.

13. For the Exchequer, the best general account remains that given by R.L. Poole, *The Exchequer in the Twelfth Century* (Oxford: Clarendon Press, 1912). More detail, including the background to the awkward 'pounds, shillings, and pence' system can be found in the author's article 'Weight, Coinage, and the Nation *c.*973–1200', *British Numismatic Journal* 83 (2013) 75–100.

14. Extracts from al-Khwārizmī's *Algebra* can be found in [K, 542–7]. See also V.J. Katz and K.H. Parshall, *Taming the Unknown: A History of Algebra from Antiquity to the Early Twentieth Century* (Princeton: Princeton University Press, 2014). In modern notation, the solution of the quadratic equation $x^2 + 10x = 39$ is given by the formula

$$\sqrt{\left(\frac{10}{2}\right)^2 + 39} - \left(\frac{10}{2}\right).$$

Nowadays we allow two values for the square root, so that there is also a negative answer (-13), as well as the answer (3) given by al-Khwārizmī.

15. For the work of Jordanus see B.B. Hughes (ed.), *Jordanus de Nemore: De Numeris Datis* (Berkeley: University of California Press, 1981).

CHAPTER 5 THE END OF THE MIDDLE AGES

1. A fascinating account of mercantile activities in medieval Europe can be found in Peter Spufford's *Power and Profit* (London: Thames and Hudson, 2002).

2. For Fibonacci, Keith Devlin's *The Man of Numbers* (New York: Walker, 2011) is recommended.

3. The solution to the version given here is nine brown birds, ten grey birds, and eleven white birds. Abu-Kamil's version of the problem of the birds is contained in the manuscript Paris B.N. Ms.4946, ff.7b-10a.

4. The division of the lira into 20 soldi and 240 denari was the same as the division of the English pound into 20 shillings and 240 pence. These systems had a common ancestor dating back to the time of Charlemagne (*c.*800).

5. The rule is: take the average of 8 and 65/8, giving the second approximation 8 1/16. The square of this differs from 65 by a small amount, 1/256, whereas the square of 8 (the first approximation) differs from 65 by 1. The beauty of the method is that the same rule can be applied again and again, giving a sequence of approximations which get better and better. This algorithm is based on the fact that if x is an approximation to $\sqrt{d}$, then (under certain conditions) a better approximation is $(x + d/x)/2$. This rule is most useful when $d = a^2 + e$ and e is small, so that the first approximation is a and the second approximation is $a + e/2a$.

6. Some specific cubic equations had been solved numerically. The most impressive result was given by Leonardo in his *Liber Abbaci*. He considered the equation that we write as $x^3 + x^2 + 10x = 20$. It is fairly easy to calculate that there is a root close to $4/3$, and Leonardo found a much better approximation, which he expressed in sexagesimal form as $[1|22|7|42]$. (This method of representing fractions was still much used for astronomical calculations, and would have been very familiar to Leonardo.)

7. Tartaglia's algorithm for the equation $x^3 + cx = d$ is as follows. First, find numbers u and v such that $u - v = d$ and $uv = (c/3)^3$. Next, find the cube roots of these numbers, so $u = p^3, v = q^3$. Then the solution of the equation is $x = p - q$. For the proof, note that $(c/3)^3 = uv = (pq)^3$, so $c = 3pq$ and $d = u - v = p^3 - q^3$. Using the binomial formula for $(p-q)^3$, and the rules of elementary algebra, we have

$$x^3 + cx = (p-q)^3 + 3pq(p-q) = p^3 - q^3 = d.$$

8. The reason why this complicated arithmetic leads to such a simple answer is that the cube roots of $8\sqrt{5}+16$ and $8\sqrt{5}-16$ are actually $\sqrt{5}+1$ and $\sqrt{5}-1$. Once again the binomial formula is crucial, because it tells us how to calculate the cube of $\sqrt{5}+1$.

(Try it.)

9. The antiquity of this branch of mathematics is attested by the fact that the problem of the sevens is remarkably similar to one that appears in the Rhind Papyrus (*c.*1650 BC).

10. In modern notation the basic property of the triangular pattern is that

$$\binom{n}{r} = \binom{n-1}{r-1} + \binom{n-1}{r}.$$

The proof depends on the combinations of n objects with r members being split into two classes: those containing a given object and those not containing that object.

11. There are many internet sites relating to cryptography, but they are very variable in quality, and it is better to rely on good books such as Simon Singh's *The Code Book* (London: Fourth Estate, 2000). The original reference to Caesar's system is in the *Lives of the Caesars*, written by Suetonius in the second century AD.

12. MATHS IS GOOD FOR YOU.

13. You may object, quite rightly, that some permutations are unsuitable for encryption, because not all letters are altered: for example, the permutation in which the letters Y and Z are swapped and the other letters are unchanged. However, a famous result in modern combinatorics gives the exact number of permutations in which all letters are changed, and that number is also huge.

14. The Arabic work on this subject is described by I.A. al-Kadi, 'The origins of cryptology: the Arab contribution', *Cryptologia* 16 (1992) 97–126.

CHAPTER 6 A NEW WORLD OF MATHEMATICS

1. Sir Walter Raleigh's royal patent can be seen at avalon.law.yale.edu/16th_century/raleigh.asp (accessed 1 June 2015).

2. The history of trigonometry is beautifully described by Glen Van Brummelen, *The Mathematics of the Heavens and the Earth* (Princeton: Princeton University Press, 2009).

3. One reason for preferring decimals to fractions is that it is easier to compare numbers. For example, would you prefer to be one of 11 people sharing an inheritance of 25000 dollars equally, or one of 9 people sharing 20000 dollars equally?

4. Harriot's papers are available as part of the European Cultural Heritage Online (ECHO) project, echo.mpiwg-berlin.mpg.de (accessed 1 June 2015).

5. A neat way to establish the properties of the curve is to consider its intersections with the line $y = tx$, for various values of t. Elementary algebra leads to simple formulas for the coordinates: $x = (6t)/(1+t^3)$ and $y = (6t^2)/(1+t^3)$. It is not known whether Descartes or Fermat used this method.

6. The terms in Archimedes' sequence of approximations 5/4, 21/16, 85/64, ..., are fractions of the form $(4m+1)/(3m+1)$. Our notation makes it easy to spot this rule, and to deduce that, as $(4m+1)/(3m+1)$ is very close to $(4m)/(3m)$, the limit is 4/3.

7. Newton's papers are available in a critical edition edited by D.T. Whiteside, *The Mathematical Papers of Isaac Newton*, 8 volumes (Cambridge: Cambridge University Press, 1967–81). His correspondence is available in H.W. Turnbull (ed.), *The Correspondence of Isaac Newton*, 7 volumes (Cambridge: Cambridge University Press, 1959–77).

8. In Newton's own formulation the symbols P and PQ were used instead of a and b. He assumed that Q is a small number, so that adding PQ to p represents a small change. He wrote

$$(P + PQ)^{m/n} = A + B + C + D + \dots,$$

in which case the rule for calculating the terms successively becomes

$$A = P^{m/n}, \quad B = \frac{m}{n}AQ, \quad C = \frac{m-n}{2n}BQ, \quad D = \frac{m-2n}{3n}CQ, \dots.$$

For example, suppose you want to use this method to find the square root of 1.2, that is, $(1+0.2)^{1/2}$. In Newton's notation, you must take $P=1, Q=0.2, m/n = 1/2$. Then $A=1$ and

$$B = \frac{m}{n}AQ = \frac{1}{2}(0.2) = 0.1,$$

$$C = \frac{m-n}{2n}BQ = \frac{-1}{4}(0.1)(0.2) = -0.005,$$

$$D = \frac{m-2n}{3n}CQ = \frac{-3}{6}(-0.005)(0.2) = 0.0005,$$

and so on. Thus $\sqrt{1.2}$ is approximately 1.0955. And we can make the approximation as accurate as we please, by calculating enough terms.

CHAPTER 7 MATHEMATICS ASCENDING

1. A good biography of Newton is R.S. Westfall's *Never at Rest* (Cambridge: Cambridge University Press, 1980).

2. Newton's work at the Mint is described in J. Craig, *Newton at the Mint* (Cambridge: Cambridge University Press, 1946), but Craig's comments on the Newton-Leibniz controversy are partisan.

3. See A.R. Hall, *Philosophers at War: The Quarrel between Newton and Leibniz* (Cambridge: Cambridge University Press, 1980).

4. Several historians of mathematics have asserted that the logarithmic property of the area under a hyperbola was discovered by St Vincent, or by his pupil de Sarasa. However, R.P.

Burn has made a careful study of the original sources and has concluded that neither of them stated the conclusion explicitly. See his paper in *Historia Mathematica* 28 (2001) 1–17.

5. Observe that $E(2) = E(1+1) = E(1)E(1)$, $E(3) = E(2+1) = E(2)E(1)$, and so on.

6. Mersenne qualified this statement with the phrase *quocumque modo hactenus cognito* (by any method so far known). In Chapter 11 we shall explain how better methods were discovered.

7. The search for Mersenne primes is now conducted by means of an internet project known as *GIMPS*, the Great Internet Mersenne Prime Search, www.mersenne.org (accessed 1 June 2015). Another project, *primegrid*, www.primegrid.com (accessed 1 June 2015), is devoted to finding primes of slightly more general kinds.

8. The fact that $2^{67} - 1$ is not prime was proved by Édouard Lucas in 1876, but the factors were not found until 1903, when Frank Nelson Cole revealed them in a talk to the American Mathematical Society. His presentation was almost entirely silent; it consisted of calculating $2^{67} - 1$ and using long multiplication to show that the product of the two factors is indeed that number. He was accorded a standing ovation.

9. Simon Singh's *Fermat's Last Theorem* (London: Fourth Estate, 1997) contains an account of Wiles' discovery.

10. Given a primitive root r, for any two numbers a and b we know that $a = r^u$ and $b = r^v$. It follows that $ab = r^{u+v}$: in other words, u is a kind of logarithm of a, and v is a kind of logarithm of b, and we can calculate the product ab by working out the sum of the logarithms, $u+v$.

11. Although the system of real numbers has the comforting property that (given the right circumstances) limits can be proved to exist, it is unfortunately not true that any set of real numbers can be measured. This fact leads to some remarkably counter-intuitive situations. For example, it is possible to divide a ball into pieces and reassemble them to make two balls of the same size as the original one.

CHAPTER 8 TAKING A CHANCE

1. A good introduction to the early history of probability and statistics is F.N. David's *Games, Gods, and Gambling* (London: Griffin, 1962).

2. *De Vetula* is a complicated allegory above unrequited love (so it is said) and the title literally means 'On the Old Woman'. The British Library copy is BL Ms. Harl. 5263.

3. The Fermat-Pascal letters are available in English, in F.N. David's book (as note 1).

4. This Nikolaus Bernoulli was the son of a brother of Jakob and Johann. There was another Nikolaus Bernoulli, son of Johann, who also made contributions to mathematics.

5. See David Bellhouse's article 'The Problem of Waldegrave', available online at www.jehps.net (accessed 1 June 2015). See also D. Bellhouse and N. Fillion, 'Le Her and other problems in probability discussed by Bernoulli, Montmort, and Waldegrave', *Statistical Science* 30 (2015) 26–39.

6. The name 'Zermelo's Theorem' occurs frequently in the current literature of Game Theory, but many of the statements attached to it are only loosely related to what Zermelo actually wrote in his article. See U. Schwalbe and P. Walker, 'Zermelo and the Early History of Game Theory', *Games and Economic Behavior* 34 (2001) 123–37.

7. See S.M. Stigler, *The History of Statistics* (Cambridge, Mass.: Harvard University Press, 1980).

8. Playfair published many similar diagrams in his *Lineal Arithmetic* (London 1798).

CHAPTER 9 MODELLING AND MEASURING

1. In Chapter 5 we saw that the area under the curve $y = 1/x$ is the hyperbolic logarithm $\log x$. The Fundamental Theorem of the Calculus tells us that if the utility U of x items is $\log x$

then the rate of change dU/dx is $1/x$. As $1/x$ decreases as x increases, U has the characteristic property of utility.

2. The rate of change of $pF(p)$ is $p(dF/dp)+F(p)$, so the maximum occurs when this is zero: that is, at the value of p which satisfies the equation $p=-F(p)/(dF/dp)$.

3. Gossen's work was published in German in 1854. An English translation is available: *The Laws of Human Relations and the Laws of Human Action Derived Therefrom* (Cambridge, Mass.: MIT Press, 1983).

4. A full discussion of Condorcet's work and its critics is contained in Duncan Black, *The Theory of Committees and Elections* (Cambridge: Cambridge University Press, 1958).

5. For Lewis Carroll, see R.J. Wilson, *Lewis Carroll in Numberland* (London: Allen Lane, 2008).

6. Ken Alder's *The Measure of All Things* (London: Little, Brown, 2002) describes the many problems of measuring the meridian.

7. The statement that the speed of light is 299792458 metres per second remains true, but for a different reason.

8. The ongoing story of the quest for standards based on the fundamental constants of physics is told in Terry Quinn's *From Artefacts to Atoms* (Oxford: Oxford University Press, 2011).

CHAPTER 10 MATHEMATICS AND MONEY IN THE AGE OF INFORMATION

1. Florence Nightingale mostly used the kind of diagram that we call a pie-chart. But she also used charts like Playfair's, for example to compare the numbers of soldiers lost to the army for various reasons.

2. See Mary S. Morgan, *The History of Econometric Ideas* (Cambridge: Cambridge University Press, 1990).

3. Bachelier's work is described by Mark Davis and Alison Etheridge, *Louis Bachelier's Theory of Speculation* (Princeton: Princeton University Press, 2006).

4. In England the banking system developed late, by comparison with some other European countries. The first English 'bankers' were London goldsmiths in the second half of the seventeenth century.

5. British Museum Online Collection, No. 1851,0901.848.

6. If each number between 1 and 8 has probability 1/8 then the expected number of questions is

$$\frac{1}{8}(1+2+3+4+5+6+7+8)=4.5.$$

7. The logarithm with respect to base 2 is related to the common logarithm (to base 10) by the rule that $\log_2 n$ is approximately equal to $3.222 \times \log_{10} n$.

8. An introduction to the mathematical theory of information, including data compression, can be found in my book: *Codes: An Introduction to Information, Communication and Cryptography* (London: Springer, 2008).

9. The stream of data that produces the compressed sequence is GGGWGGGGGGWGGGGW GGGGWGGGGGGGGGWWGGGGGGGGWGGGGGGGWGGGGGGGGGGGWGWGGGGGW.

CHAPTER 11 CAN MATHEMATICS KEEP US SAFE?

1. See Simon Singh, *The Code Book* (London: Fourth Estate, 1999).

2. The algorithm of Agrawal, Keyal, and Saxena is based $(x+y)^p$ being congruent to x^p+y^p modulo p, a result that was first observed by Leibniz.

3. Martin Gardner's article on RSA is 'A new kind of cipher that would take millions of years to break', *Scientific American* 237 (August 1977) 120–4.

4. It is now known that a system very similar to RSA had been suggested a few years earlier by mathematicians working for GCHQ, the British Cryptographic Service, but this fact was not made public until 1997. See *The Code Book* (as note 1).

5. The algorithms for encryption and decryption in RSA both operate on messages expressed in the form of integers modulo n. An original message m is encrypted as m^e, and an encrypted message c is decrypted as c^d. The system works because e and d are chosen so that $m^{ed} = m$ (modulo n).

6. This is one of the seven 'Millennium Problems' for which the Clay Mathematical Institute has offered a prize of one million dollars.

7. There are many online references to the RSA Factoring Challenge, for example www.emc. com (accessed 1 June 2015).

INDEX